WOMEN'S VOICES FROM THE RAINFOREST

The rainforest has become increasingly topical in today's eco-conscious society, yet people remain ignorant of the many issues concealed by the language and methods of international development policy. *Women's Voices from the Rainforest* analyses the causes and effects of such policy.

Concentrating on the women pioneers of poorer countries, the book employs a variety of contrasting methodologies, from life histories to questionnaire surveys, to suggest a range of answers to this increasing global concern.

In the first study of its kind to focus on the women in the families which are cutting down the rainforest of Latin America, the authors argue for appropriate planning which responds to local cultures, not to Western ideologies, and to which women and men contribute at all levels.

Women's Voices from the Rainforest is an analysis not only of the many costly eco-failures when 'new' lands are settled in poorer countries today, but also a discussion of the role feminist theory has to play in finding solutions to such problems.

Janet Gabriel Townsend is a lecturer in Geography at the University of Durham and wrote *Women's Voices from the Rainforest* in collaboration with Ursula Arrevillaga, Jennie Bain, Socorro Cancino, Susan F. Frenk, Silvana Pacheco and Elia Pérez.

INTERNATIONAL STUDIES OF WOMEN AND PLACE

Edited by Janet Momsen, *University of California at Davis* and Janice Monk, *University of Arizona*

The Routledge series of *International Studies of Women and Place* describes the diversity and complexity of women's experience around the world, working across different geographies to explore the processes which underlie the construction of gender and the life-worlds of women.

Other titles in this series:

DIFFERENT PLACES, DIFFERENT VOICES
Gender and development in Africa, Asia and Latin America
Edited by Janet H. Momsen and Vivian Kinnaird

'VIVA'
Women and popular protest in Latin America
Edited by Sarah A. Radcliffe and Sallie Westwood

FULL CIRCLES
Geographies of women over the life course
Edited by Cindi Katz and Janice Monk

SERVICING THE MIDDLE CLASSES
Class, gender and waged domestic labour in contemporary Britain
Nicky Gregson and Michelle Lowe

GENDER, WORK, AND SPACE
Susan Hanson and Geraldine Pratt

WOMEN'S VOICES FROM THE RAINFOREST

Janet Gabriel Townsend

in collaboration with
*Ursula Arrevillaga, Jennie Bain,
Socorro Cancino, Susan F. Frenk,
Silvana Pacheco and Elia Pérez*

ROUTLEDGE

London and New York

First published 1995
by Routledge
11 New Fetter Lane, London EC4P 4EE

Simultaneously published in the USA and Canada
by Routledge
29 West 35th Street, New York, NY 10001

Typeset in Baskerville by
J&L Composition Ltd, Filey, North Yorkshire
Printed and bound in Great Britain by
Biddles Ltd, Guildford and King's Lynn

British Library Cataloguing in Publication Data
A catalogue record for this book is available from the British Library

Library of Congress Cataloging in Publication Data
Townsend, Janet G.
Women's voices from the rainforest/Janet Gabriel Townsend.
p.   cm. – (International studies of women and place)
Includes bibliographical references (p.   ) and index.
1. Women in development–Environmental aspects–South America.
2. Women in development–Environmental aspects–Tropics.
3. Rain forests–South America. I. Title. II. Series.
GV1547.T66 1994
793.8–dc20   94–12466

ISBN 0–415–10531–5
0–415–10532–3 (pbk)

CONTENTS

Acknowledgements   vii

PREFACE: THREE CONVERSIONS   1
Janet Townsend

Part 1 Outsiders

1 INTRODUCTION: WHO IS AN EXPERT?   7

2 WOMEN PIONEERS IN THE TROPICS   18

3 IN THE COLOMBIAN RAINFORESTS   34

4 WOMEN PIONEERS IN MEXICO: OUR ANALYSIS   50

Part 2 Outsiders and insiders

5 MEXICAN WOMEN PIONEERS TELL THEIR STORIES   81

6 OUTSIDERS' CONCLUSIONS   123

Part 3 Insiders' voices? Mexican women speak

7 RE-PRESENTING VOICES: WHAT'S WRONG WITH
  OUR LIFE HISTORIES?   137
  Susan F. Frenk

8 CARMELA'S LIFE STORY   144

9 ELENA'S LIFE STORY   154

10 CLARA'S LIFE STORY   167

11 GUADALUPE'S LIFE STORY   180

Bibliography   194
Index   206

ACKNOWLEDGEMENTS

Our thanks are due in the United Kingdom to the Economic and Social Research Council and to Durham University, who funded the research, and in Colombia to the Peasant Action Association and the Ministry of Agriculture, Irrigation Section. We owe to Donny Meertens (Santafé de Bogotá, Colombia) the idea of seeking women's own solutions. In Mexico we thank the Colegio de Michoacán, especially Dr Birgita Boehm de Lameiras and Dr Gabriel Muro, who provided support and secretarial services. Also in Mexico, we learned much from Dr María Elena Alvarez-Buylla Roces, who helped us with forest gardens, and from Dr Lourdes Arizpe and Dr Magalí Daltabuit, who shared with us their projects in the Lacandón rainforest. Since we could not return to these remote, scattered communities to discuss our conclusions and check life stories, the names of women and communities where we recorded life stories are fictional. We are sorry that we cannot blazon their contribution by name: we owe them an immense debt. In England and the United States, we thank Hope Page, Alan Townsend and above all Janice Monk and Janet Momsen for their great improvements to the text.

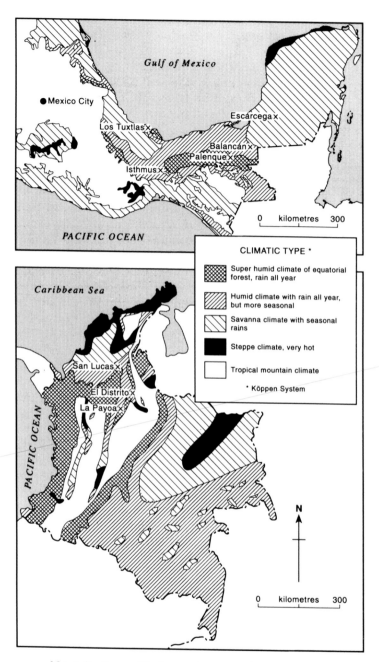

CLIMATIC TYPE *

Super humid climate of equatorial forest, rain all year

Humid climate with rain all year, but more seasonal

Savanna climate with seasonal rains

Steppe climate, very hot

Tropical mountain climate

* Köppen System

Map 1 South-east Mexico: study communities and climates

Map 2 Colombia: study communities and climates

PREFACE: THREE CONVERSIONS

Janet Townsend

The voices which we want to speak in this book are those of women pioneers. To help the reader hear the voices, it is important to know how the research came about, and how it was motivated. As the instigator of the work, I must explain myself. Since 1965, I have been involved in research with people trying to make farms out of the rainforest in Latin America – the pioneers on the frontier of colonisation, or the people blamed for deforestation, according to one's point of view. Working with people who are fighting to make a sustainable livelihood for themselves and their children, often at the cost of great suffering and risk, has always been very depressing. In Latin America, most of these pioneers fail and lose their farms, sometimes their lives. Much land goes from forest into non-sustainable cropping, and eventually to non-sustainable cattle ranching. Environmental degradation, human impoverishment and failure are normal.

I undertook research for my doctoral thesis in the middle Magdalena valley, Colombia, which has been transformed while I have known it from rainforest to farms and then to ranches with low levels of productivity, wages and employment. The pioneers I knew in the 1960s have gone, and the profits from these ranches now go to faraway land speculators and urban banks. Recently, the most profitable land uses in the valley have been airstrips and processing laboratories for cocaine. In the 1960s the army terrorised the countryside, but today the great fear is of the paramilitary death squads, followed by the cocaine dealers, the kidnappers and lastly the army and the guerrillas. Living standards are higher, but life is more insecure than ever.

I have continued my work on colonisation not because I see it as a good source of sustainable livelihoods, but because the pioneers have worked so hard to educate me. During the years since 1965 my perspectives on their problems have shifted radically at least three times.

FIRST CONVERSION: FARMERS AS SCIENTISTS

I began my doctoral research with a long-standing conviction that my own culture had produced rather a poor design for living. We had an unsatisfactory,

1

unenviable society, but had made certain technical advances which could, I thought, be beneficial to other societies, who might use them to make a better job of living than we had done. I belonged to the long-lived 'transfer of technology' school of which writers like Robert Chambers *et al.* (1989) are so critical. This story is prehistory – I had not heard of social change, of dependency theory or of political economy, and my supervisor regarded postgraduate training as an undesirable waste of time. I was also under the illusion that intensive research was not really geography, and my work was both gender-blind and male-biased.

As a doctoral research student, I was sent from Oxford to Colombia from 1966 to 1968 to study the 'ignorant peasants' who had been displaced from the highlands by rural violence and were destroying the forest resources of the middle Magdalena (Townsend 1976). Unfortunately, it took me much of this time to realise how completely I had been misdirected and to undergo my first conversion. I found, of course, capable people making remarkable adaptations to their new environment, but who were at the mercy of the political economy of Colombia. It took me a long time to realise that there was no known technical solution for their problems and that the agronomists, hydrologists and forestry experts who pontificated on the ignorance of the settlers could offer no viable alternative farming, fishing or forestry systems.

In the late 1960s, the settlers in the middle Magdalena lived in fear in a situation of terrifying social conflict, but had great faith in the power of the written word and of experts. They set a high value on my level of education, for they saw knowledge as power and believed in the value of truth. Like me, the settlers believed that if they could get their story told, it would make a difference to their problems. However, no one in the Colombian establishment, whether in the Magdalena Valley Corporation, the Ministry of Agriculture or academic geography, would believe me. My methodology was based on unstructured interviews, not questionnaires, and so was regarded as subjective and unscientific. I was told that I had 'imposed my perceptions on my respondents', that the respondents were 'stupid, traditional, ignorant and destructive', and that the skills and eagerness to learn which I described were 'figments of my imagination'.

Returning to the area in 1974, I used psychological tests and computers (with repertory grids) to demonstrate that the settlers' ideas about good land use were similar to those of their agricultural advisers, but that the settlers did not have the resources to put their ideas into practice (Townsend 1977). To my surprise, this work also failed to convince anyone. What I had not realised, in all this time, was how convenient these beliefs were to the Ministry of Land Reform, to academics and to experts, and therefore 'proof', with or without computers, was irrelevant. The settlers, like so many of the poor and powerless, were very effectively denied a voice, for legitimate knowledge was the prerogative of the elite.

SECOND CONVERSION: WOMEN'S VOICES

In 1984, I began to study the plight of the women among the pioneers, and this led to my second conversion. My first conversion had shown me how poor settlers were silenced, devalued and disqualified. Now I learned how women's work was rendered invisible. I had of course talked with women pioneers a great deal before, when I had stayed in their houses, camped near their houses and talked all night by the cooking fire. For these conversations I had used very loosely structured interviews, open conversation and, once, repertory grid techniques. My primary informant had usually been the man of the house, although sometimes I interviewed the whole family. Indeed, on several occasions, my research assistant had entertained the women and children, so that the man might concentrate on the repertory grid! I had always had an androcentric agenda, a male-centred agenda, for I had been interested in the experiences and prospects of the farming household, and even when chatting with the women, I had never enquired about their personal lives as they did about mine, for it seemed intrusive.

When, in 1984, there was an amnesty and I was able to return to the Magdalena valley with Sally Wilson de Acosta, I discovered that many of my assumptions about the women pioneers were quite wrong (Chapter Three). As a result, we tried to make these invisible women visible, to recount their strengths and their suffering (Townsend and Wilson de Acosta 1987) through a questionnaire survey and interviews. However, there was no official response, despite our reports and article. In exactly the same way as the male settlers with whom I had worked earlier were denied a voice, so voices raised on behalf of the women were skilfully disqualified.

I shall never forget one interview with a woman deep in the forest. She was very nervous and, as her two undernourished children cowered in the kitchen, I tried to complete my questionnaire quickly and leave her in peace. As I headed down the path, she called, 'Oh, please come back! I don't know why I was scared, but I haven't seen anyone but him and the children for months!' As we chatted for a while, I felt even more conscious of privilege than usual.

After these experiences in Colombia, a literature survey (Townsend 1991) showed me that women around the world tended to have some quite similar experiences in colonisation or land settlement. If the colonisation was planned, women were often left out of the plans, for planners thought only of moving 'men without lands to lands without men' (Chapter Two).

But what is the point of documenting more and more bad experiences which people have in common? My third shift of perspective came in 1990, when I tried in Mexico, with Jennie Bain, to give pioneer women more control over the agenda. We used the same questionnaire but when we began to learn to use life histories, I found that I still had a male agenda. I had still been blind to women's central concerns for I had still left the personal out of account.

THIRD CONVERSION: THE PERSONAL IS POLITICAL

The aim of the Mexico project, in 1990 and 1991, was to develop with pioneer women some guidelines for women's grass-roots organisations in areas of land settlement in Mexico, and perhaps some guidelines for the planners (Chapter Six). I certainly found out something about my own prejudices. As I have said, I came to the women in these localities with a male agenda. My concern was with sustainable livelihoods, differentiation, sources of income, education, services. I habitually talked at the household level and had a socialist–feminist preoccupation with production. (I was fascinated by the possibilities of home gardens for sustainable agriculture, for instance.) When we asked women to talk about their lives so that we could better understand what it was to be a woman in their community, the focus shifted to very familiar Western, urban, feminist concerns with women's control over their own bodies, with child abuse and marital rape – issues which we, the outsiders, have had no more success in dealing with than they have.

This is the essence of what we will present in this book: the difference between outsiders' writings and insiders' voices, one reflection of the 'etic' and the 'emic', the 'facts for the analyst' and the 'facts for the subject'. We shall try to situate this knowledge, with all its flaws. For instance, we have not taken the book back to all our subjects or even to all our tellers of life stories. The information and life stories obtained were explicitly given to be published, but would the givers now approve of the extracts which we are publishing? We cannot know. For this reason, the people and communities who gave us life stories and allowed us to hold workshops are identified only by fictional names.

Part 1

OUTSIDERS

1

INTRODUCTION: WHO IS AN EXPERT?

If we don't cut down the forest, what will there be to eat?

(Cristina, Cuauhtémoc, Mexico, 1990)

This book sets out to explore two questions: the position of women pioneers in the settlement of new lands and the methodological difficulties of such a study. We shall therefore describe different ways of understanding the problems of women pioneers. We use a literature review (Chapter Two), questionnaire surveys (Chapters Three and Four) and life stories (Chapter Five and Seven to Eleven), in order to give different perspectives both on the women's lives and on the advantages and disadvantages of various research methods and objectives. In Chapter Six, we report on the position of women pioneers and seek solutions.

Pioneers, settlers, homesteaders, colonists: these are all evocative names for people who set out to make farms on land where little or nothing was produced before, be it on a new irrigation scheme or in the wilderness. Sometimes such a movement of people is planned, sometimes spontaneous. Migration to the frontier of settlement has had different names at different times and places: pioneering in the USA, land settlement in Africa and South Asia, colonisation in Latin America, transmigration in Indonesia. Pioneering is not as simple as it sounds, for pioneers are men, women and children who think they are moving to 'empty' lands but often find themselves displacing the previous inhabitants, whether American Indians in the USA, Aborigines in Australia, tribal peoples in India or Amerindians in Brazil. The World Bank calls these displaced people 'oustees'. Many pioneers are themselves 'oustees'.

WOMEN PIONEERS

Why study women pioneers? A hundred years ago, pioneers were heroes and heroines opening up new lands and creating new wealth, in the USA, Australia, Russia and the European empires. Now, 'pioneers' are moving into much more fragile environments and the inhabitants they seek to displace are a little more visible to the outside world, so that television viewers are more likely to see most

7

pioneers as enemies of the human race, destroying indigenous peoples or exterminating valuable species of plants and animals.

However, pioneers are important because so many governments still see land settlement as a political answer to rural poverty, particularly for people displaced by big dams (for example, China and India), by agricultural change (Brazil) or by conservationists (proposed in Cameroon). We agree with David Hulme (1987) that, although very few land settlement schemes ever achieve their objectives (these few being very costly), land settlement continues to appeal to governments, to bureaucrats, to commercial interests and to international funding agencies. Further, we have no wish to see more land settlement, as other options usually provide higher economic returns overall, although profits for individual politicians, officials and speculators may be less.

The conditions for successful spontaneous pioneering now exist in very few places, while what planned pioneering needs, but almost never gets, is excellent administration under very difficult conditions. But millions of people are already pioneers, often living in terrible conditions and rarely earning a sustainable livelihood. Unless solutions are found, many will be forced to move to more 'new' lands at the cost of even more destruction for little gain to themselves.

Thus, pioneers still matter. Why separate women pioneers? We shall argue that women often suffer more than men as pioneers, that women often have different interests and needs from men and may identify different problems and propose different solutions. Women have stories to tell.

Women pioneers in this book

Who is the expert? Women have stories to tell, each different, while our analyses of their situations may be different again and we may well disagree with each other. Identifying the expert is a fundamental problem in social science. In Part 1, the experts are outsiders, specialists, who are not members of the community. Chapters Two and Three will present the academic as the expert who seeks to be objective, rational and detached in order to establish the truth: the outsider's view, the analytical view (Merton 1972). In Chapters Seven to Ten the experts will be insiders, pioneer women giving their own testimonies, experts in their own lives. This sequence repeats the story of this research.

In a sense, the book moves from the 'etic' to the 'emic', from the outsider's analysis from outside the system to the insider's understanding from within the system (Merton 1972). First, in Chapter Two, we shall illustrate the distant, authoritative, academic, outsiders' view of women pioneers, from a survey of published work. We shall show that, although there is a massive literature on land settlement, most of it is written as if there were no women involved, so that the millions of women pioneers are invisible. Using a handful of studies of women pioneers around the world as a basis, we shall argue that women often suffer more than men as pioneers. Planned colonisation hardly ever takes

8

account of women. Women tend to have to work harder and longer in worse conditions, to lose their rights in land, to lose control over income, to have poorer access to health care, to education and to places to buy goods, and to suffer even more painful isolation than the men. There are great differences, of course, between different places and cultures. The surprise is that there is similarity in findings for countries as diverse as Colombia, Sri Lanka, Nigeria, Malaysia, or India and Brazil, perhaps because the planners around the world have had the same training, read the same books and make the same mistakes.

Chapter Two is very much traditional Western scholarship, providing the opportunity for outsiders to review the literature and compare findings across space and time, across thousands of kilometres and many years, as if by the application of scientific method it were possible to detach ourselves from our own place and time and make objective comparisons – as if we could have God's eye view and write from a standpoint outside human society (Haraway 1988). Archimedes said, 'Give me a lever and a place to stand, and I will move the world.' We forget how much our theories as well as ourselves are strictly products of our time and place, as we do not stand outside the world.

Chapter Three will explore diversity and difference at a much more local scale, among women pioneers in one quite small area of tropical rainforest in Colombia in 1984 and 1987. In each of three places studied there is great diversity among the women, and the differences between the places (a remote frontier area, a high-tech project and a co-operative) also create differences in the expression of gender in each location. This field research was also designed for us, as outsiders, to build up evidence, to carry conviction, to be 'objective'. Some surprises, as we shall see, came out of the computer, but this was still 'extractive research' (Nelson and Wright 1994), mining and shaping information and taking it away, centralising it for the powerful and the privileged and for the financial profit, the careers, of the researchers, the outsiders. We did return reports to the communities and, at their request, to the appropriate ministries and officials, but without producing any of the results the pioneers wanted. We published evidence of the invisibility of pioneer women in English and Spanish (Townsend and Wilson de Acosta 1987; Townsend 1989), but could do nothing to explain that invisibility or to change it.

What is achieved by demonstrating a problem? Women's particular difficulties in land settlement had been identified by Robert Chambers (1969) in what became a standard text for planners, yet apparently made no difference to pioneer women. Many development 'experts' used to believe that the truth would make a difference, as did many poor people who co-operated in development research. Most social scientists were concerned to establish and demonstrate the 'truth', to make generalisations, to look for causes and effects and often to rely heavily on quantitative data. Good methods would ensure sound results, even if all the researchers came from another culture altogether. Field

9

research was important, but only outsiders, educated in the Western tradition, were competent to evaluate or understand it ('avoid the concrete peasant', as one investigator said to an academic audience). Peasants, blinded by their circumstances and limited education, could not recognise the deep structures which controlled their lives. 'Experts' like us had to do that on their behalf.

In Colombia we tried to ask women pioneers what they saw as their problems and what solutions they thought there might be. These were unfamiliar thoughts, questions brought in by outsiders to women who saw few outsiders, let alone peculiar-looking Europeans. Also, our preoccupations were very much at the level of the household. Janet Townsend had a very socialist-feminist preoccupation with production; Sally Wilson de Acosta a professional concern with child nutrition. Knowing the extent to which poor rural women in many underdeveloped countries work on the land, and the degree to which they had been shown to do so in other areas of Colombia, we were astounded to find that these pioneer women had become housewives in the rainforest, working long hours caring for men and children, leaving the house only to do the laundry. In interviews, these women talked to us very much at the level of the household and about the needs of their families for income and for services. Although they asked us personal questions, as outsiders, intruders, guests, we felt that it would be impertinent to reply in kind. As a result, we learned little of the women's own wants and needs for themselves.

Part 2 represents our efforts to find out about and report on the views of pioneer women in Mexico in 1990 and 1991. Ursula Arrevillaga, Jennie Bain, Socorro Cancino, Silvana Pacheco, Elia Pérez and Janet Townsend used questionnaires, interviews, workshops and life stories to try to hear more of the insiders' voices. In Chapter Four, we try to represent an orthodox analysis, an etic view of women pioneers as outsiders see them, mainly through the use of questionnaires and interviews. In Chapter Five, we use their own words, mainly from life stories, to describe their lives, trying to empathise with and chronicle their experience. However, this is still an etic perspective, an outsiders' search for understanding. In Chapter Six, we set out their problems and solutions as we see them and, we hope, as they see them and we make policy recommendations for land settlement in Mexico and around the world.

In Parts 1 and 2 we seek to set out a single narrative, a single vision collectively agreed among the authors. Part 3 seeks to present insiders' voices, fragments of the emic perspective, with a selection of life stories from Mexican women pioneers, testimonies of 'what it is to be a woman in this community'. One objective of Parts 2 and 3 is 'transculturation'; at its simplest an exploration between cultures. Transculturation requires translation to include not only the words but the cultural context (see Chapter Seven), and for 'silenced voices' such as women's to be recorded and published (Basso 1990). Our account of the cultural context in Part 2 is therefore important to reading Part 3. We apologise to interested feminists for not also analysing the texts of the dominant groups who, in this case, control land settlement (compare Pratt 1992). This would be another project.

Unequal power relations, national and international, have enabled us to write about women pioneers in this way, to centralise, control and manipulate their information (Madge 1993). Ours is not collective work with the pioneers nor a testimonial process generated from the bottom up by the people themselves, for they did not choose us, nor select the life stories to represent their lives nor confirm, edit or translate them. We remain in command, the authority is ours and we must not dissemble this (Stacey 1988). We worked as a team, but our storytellers did not, for we worked with each of them individually. The whole book represents the limitations of a certain kind of research in which one university academic has access to the world literature and reads about world land settlement, about Colombia and Mexico, from distant Britain. At intervals, she gets a government grant and uses research leave to conduct some fieldwork. As this grant funded the co-authors, at least this book is not a single, external view, for several women do the work together, collectively discussing, debating, reviewing and seeking always to listen to the subjects of the research. But all the visits were short and no one engaged in participant observation for over a year or more. Even the life stories are fragments, told over an hour or two, not built up over a year or more as is usual. We shall argue that there are some benefits to this 'gender tourism' (as with Robert Chambers' 'rural development tourism' (1984)) but there are certainly grave disadvantages.

WHO ARE WE, THE AUTHORS, THE OUTSIDERS?

In the first place, we are all women and, in the more personal discussions later in the book (Chapters Five and Eight to Eleven), we worked only with women. Everyone involved in asking for, recording, translating and editing the life stories and workshops reported from Chapter Four onwards is a woman. We are all, in different ways, feminists with a deep concern and respect for the powerless and exploited, beliefs which are very important not only in shaping the research goals and methods but for personal working styles. Six of us are socialists.

In some ways, pioneer women see us, as women, as being closer to them than their own men, yet they also set us in a frame which says 'outsider'. The process of talking to us is very complicated. Each narrator of a life story was asked, 'Please tell me about the story of your life, so that I can understand what it is to be a woman in this community, and can write about it for other people.' Some women really spoke to 'other people', but each also told her story to another actual woman, an outsider who belongs to a different place, a different class, perhaps a different 'race'.

Some of us could pass physically for strangers from the next village, but four could not. In Colombia and Mexico, 'race' in terms of physical appearance is less important than in North America or Europe. 'Race' is more social and the categories much more complex. For example, in the middle Magdalena valley, most people have European, Amerindian and African characteristics, but class and region of birth are much more important than physiognomy. In Mexico, indigenous people are the poorest and most disadvantaged. Somehow, we need to 'include our own social location into the interpretation of our work' (McDowell 1992).

To give the reader a feel for our locations and our diversity as authors and interviewers, we introduce ourselves here, emphasising some of the reasons why we all represent the intractable outside world in these areas – colouring, accent, qualifications, childlessness. (Childless grown women who have no male partners with them are very strange indeed, blue eyes are quite strange, brown hair is unusual and fair hair almost unknown.) Our personal characteristics, of course, coloured not only how we were seen but what we saw.

Janet Townsend (born UK 1944, D.Phil., living in an ex-mining village near Durham) is the continuity girl in this book, linking the different chapters across a decade of research. She is blue-eyed, brown-haired, childless, plump, prosperous and instantly identifiable as foreign by all the subjects of this book. She speaks Colombian Spanish, erratically. She teaches geography at Durham, a British university, and does research in Latin America when she can (having spent a total of four years there so far). She participated in the projects we shall describe in Colombia and Mexico and raised the money for them.

Sally Wilson de Acosta (born UK 1954, MSc., two children, living in Santafé de Bogotá, Colombia) is a nutritionist with a Durham degree in geography. She has fair skin and mid-brown hair but speaks Colombian Spanish. She is not an author of the book but before her children were born, in 1984, she worked on the project in the Serranía de San Lucas, Colombia, (Chapter Three) contributing to the cumulative fieldwork. She works part time with Save the Children and UNICEF.

Jennie Bain de Corcuera (born Colombia 1957, MA, two children, living in Mexico City) grew up in Mexico but is very fair and often taken for a rich foreigner, her parents being prosperous immigrants from Switzerland and Grenada. She was a biologist interested in resource management before she decided to work at Durham on rural Mexican women and the environment. In 1990, she conducted interviews on the pilot survey in Los Tuxtlas and in Uxpanapa, Mexico, to her children's annoyance. She teaches and researches in Mexico City.

Four more of us worked collectively with Janet Townsend in 1991, conducting a survey in Mexico, running workshops, recording and transcribing life stories, writing the reports (Chapter Four) and writing a book, *Voces Femeninas de las Selvas*, for publication in Mexico (Townsend et al. 1994). Only Silvana Pacheco had studied feminism before this research. Like many Mexican social scientists, we were all trained in mildly Marxist social theory and see the fulfilment of the individual in co-operation rather than competition. We were taught, for instance, to approve the collective ownership of land which was government policy in Mexico from 1917 to 1991. We are all childless.

Ursula Arrevillaga de Escobar (born Mexico 1968, BA, living in Comitán, a town in southern, highland Mexico) is a young sociologist from the hot Pacific lowlands near the Guatemalan border. She too has mid-brown hair and her parents are fairly prosperous immigrants from Spain. Ursula had recorded the life stories of activist urban women for her degree. She now teaches in a private school.

Socorro Cancino de Córdoba (born Mexico 1958, BA, also living in Comitán) is a sociologist from Comitán and the only one of us to look not only Mexican but indigenous. She had no experience in gender research, having worked with trades unions in Mexico City and with non-governmental organisations (NGOs) in rural projects. She now works with Guatemalan refugees in the rainforest.

Silvana Pacheco Bonfil (born Mexico 1958, BA, living in Texcoco, near Mexico City) is a socio-agronomist from a village in central Mexico where her parents were peasants. She looks Spanish. Trained by Emma Zapata (see p. 128), she has worked with women in rural projects and has travelled with a scholarship to Israel and Europe. She now works in rural development.

Elia Pérez Nasser (born Mexico 1959, BA, living in Mexico City) is also a socio-agronomist. She grew up on the hot lowlands near the Gulf of Mexico but studied in Mexico City. Although her grandfather was Egyptian, she passes for Mexican. She had worked in the regulation of agricultural chemicals, though had never studied or worked with women, and she now freelances in environmental research from a base in Mexico City.

Sue Frenk (born UK 1958, Ph.D., two children, living in Durham) supported the project from England. She specialises in cultural politics and in Latin American women's narrative, especially in Mexico, which she knows well. Her contribution to this project has been through translation, discussion and editing, in England. She teaches Latin American and Spanish literary and cultural studies at Durham.

As a team, we bring diverse values and interests to the project. The agronomists had a better grasp of technical possibilities, the geographer more interest in environmental issues, the sociologists a strong interest in the 'causes' of, for instance, the vulnerability of some women to marital violence, the linguist a concern with deconstruction. The Mexicans seek explanation and prediction from social studies, while the British authors seek to learn from subaltern voices (Chapter Seven). But we are all in many ways outsiders. None of us has ever been a pioneer, though Silvana grew up as a peasant. Nearly all of us live in towns and cities. All of us have academic training. None of us is lesbian, and we learned nothing of lesbian lives; all but one of us had male partners at the time of the interviewing. We are friendly strangers, not participant observers, for our encounters and even our stays in given areas were short. To pioneers, we are the elite: educated, powerful, with good access to information, bureaucrats, politicians. The outsider/insider axis (Jameson 1982) is of course a continuum. As women, we are all insiders, yet Janet Townsend is foreign, Jennie Bain (Mexican) looks even more foreign, while Socorro Cancino looks very Mexican, and our perspectives and goals are also diverse. We shall indicate in the course of the book how our differences were expressed. We hope that this book incorporates all our insights, but it is important to remember that we all represent the 'outside world' in these communities.

WHO ARE YOU, THE READERS? WHAT'S IN IT FOR YOU?

We are seeking to represent pioneer women to specialists in 'development' and above all to the planners of land settlement. We want to show that the discourse which renders them invisible also harms pioneer men, and often the project itself. We want, above all, to see that discourse change. To many such specialists, the language of discourse, text, context or representation is unfamiliar, as are feminisms and feminist theories, so we shall avoid such language as far as possible.

Nevertheless, we hope that feminists will be interested at different levels in the tales of pioneer women, in the history of their invisibility and the narratives elicited by the different methodologies. We shall try to give pointers to relate our narrative to feminist debates and shall cross-reference more than usual, for those not reading all the book.

Many feminists will be interested mainly in Chapters Seven to Eleven where we discuss the difficulties of transculturation and seek to present the voices of pioneer women. Nevertheless, awareness of context is of supreme importance to any success in transculturation and in this Chapter Five and even Chapter Four will be helpful.

Some students of 'development' may be more interested in Chapter Two, where we review briefly the experiences of women around the world, and in Chapter Six, where we discuss 'solutions'.

FEMINIST METHODS?

We shall recount our methodology in detail at relevant points in the book, but we need also to ask, 'what can we, as outsiders, contribute?' As scholars, we want to know, to understand. As feminists, we want to represent the lives of pioneer women and to give them some opportunity to represent themselves, with their problems and their solutions, while recognising that only partial success in these aims is possible. We do not think that such representation offers any universal answer to the problems of land settlement, only that few men pioneers have ever been asked to suggest solutions and even fewer women. There will be no solution without these voices.

Because we want these pioneer women to represent themselves, we are entangled at once in the difficulties of representation. We are working towards what Donna Haraway (1991: 190) has called 'feminist versions of objectivity'. We admit that the knowledge we are presenting is incomplete for it is limited by our own personalities, understandings, cultures, languages, ethnicities, classes and perhaps even gender. It is partial in two senses, for it is incomplete and it is committed. We want to situate this knowledge, to be explicit about its creation and origins, to be sensitive to the structures of power in which we and our subjects are acting. We are committed to making visible the claims of the less powerful (McDowell 1992), but how can we do that?

It used to seem a very attractive project for insiders, the subjects of research, to control the agenda of research and for academics to represent them. In a sense, this has been a traditional role of the intellectual in Latin America, as spokesperson for the voiceless (Yúdice 1991). We now have to recognise that when academics represent people, we do not display them through a clear sheet of glass but rather re-present them in our own terms (James Clifford 1986, would say we construct them and/or invent them).

Literature review

In reading published work, academics in core countries have the power as never before to find and read what exists. (For example, Janet Townsend, working at a British university, has excellent bibliographical search facilities and abundant access to interlibrary loans.) However, the centralisation of knowledge has greatly harmed many academics and non-academics in the periphery.

Interviewing

In interviewing, feminists want to develop less exploitative and more egalitarian relationships. Anne-Marie Goetz (1991) and Stephen Pile (1991) have proposed that we can only do this through creating alliances between researchers and subjects. Such alliances may be open to researchers who follow Maria Mies' (1983) prescription to work only when asked to do so by an existing, politically active group, but can it be applied in our case?

Think of the context of our research. We arrive in command of a means of transport (mule, jeep or horse in Colombia, public transport, pick-ups or horse in Mexico). The transport defines us as in command of resources, which is emphasised by our colouring, speech, education, extraordinary presence in the area and improbable explanation of ourselves. In Colombia, we speak of learning in order to teach back at home, while in Mexico political talk, at least, has been socialist for many years, so that people understand when we call ourselves 'comrades' and 'sisters' concerned for the welfare of the community. (What they believe is another matter.) We are well-fed and healthy so that even our bodies tell of power and prosperity (Patai 1991). We come with specific

15

introductions – from a peasant league or a World Bank project in Colombia, from a Mexican university and local officials in Mexico. In some places we are guests, slinging our hammocks to sleep, but in others we visit daily, never for more than a fortnight. We are at best 'friendly strangers' (Cotterill 1992). We are an entertaining circus, the foreigner among us even more so. We represent power. A decade ago (Mies 1983) it seemed that involvement and interpretative methods could stop us using people as mere objects of research, but now Judith Stacey (1988) has shown that the subjects of research may be even more vulnerable when personal links are stronger, for these very links place them at grave risk of manipulation and betrayal: 'For no matter how welcome, even enjoyable, the field-worker's presence may appear to "natives", fieldwork represents an intrusion and intervention into a system of relationships [. . .][1] The inequality and potential treacherousness of this relationship is inescapable' (Stacey 1988: 23). We tried to be very open about our objectives, and our subjects are protected a little by the shortness of our stay, but, as Daphne Patai writes (1991), the ideas of sisterhood, of being women together, the 'shared space of femaleness' (Jameson 1982), give us access but are in many ways a fraud, for we are still extracting material from our subjects for our use – exactly what Ann Oakley objected to in her classic contribution (1981). We were accepted for our gender, our politics and our personal styles but we agree with Linda McDowell (1992: 408) that 'The notion of non-exploitative research relations is a utopian ideal that is receding from our grasp.'

We have tried to return the results of the research. We have sent reports to the communities and to interested and literate respondents. We also sent reports, as asked, to appropriate ministries and individuals in power, although our subjects know as well as we do that the chance of a positive response is small. We wrote up all workshops and returned the originals to the groups, but we recognise that following these feminist precepts may be little more than 'feelgood' measures for us (Patai 1991). We do not think that people gain enough from being interviewed or telling their life stories for these benefits to justify the process. Women welcomed us, once we had persuaded them not to be afraid, and responded to interest very positively. We did pay them in kind for their time as a mark of respect and we tried to make the interview a good experience. To some degree, personal style becomes more important than initial appearance and difference as the work progresses. We were always conscious of our privilege and power, for, materially and personally, we gain so much more than they. We tried hard not to impose our perceptions upon them (Ong 1988). We sought to be open, explicit and non-exploitative but we cannot know the extent to which we failed.

Life stories

The telling of life stories is a kind of negotiation, a matter not only of the analyst trying to place the women but of each woman trying to place the analyst

(Phillips 1990) and of the creative tension between advocacy and scholarship (Gluck 1991). In addition, most women seek to manipulate the analyst and probably succeed. Both parties have an agenda. One weakness of the feminist debates about interviewing is that they tend to imply passivity and even incompetence on the part of the subject. Claudia Salazar (1992) gives powerful insights into the cultural politics around *I, Rigoberta Menchu* (translated by Elisabeth Burgos-Debray 1984) but is critical: 'Rigoberta's story is insured a place in bookstore shelves for the "facile consumption of cultural Otherness".' There is an implication that Rigoberta has been made by the West into an object for consumption without her realising it, but Rigoberta is a Guatemalan Indian, and no Guatemalan Indian woman could be unaware of being made into an object of consumption every day for tourists, or be in doubt of what would happen to her narrative. Rigoberta's political thought may not be Western but it is powerful and highly developed and we think that even at age 23 she made a political calculation in telling her story. For all her ignorance of Western politics, she was to win a Nobel Peace Prize.

As we shall describe in Chapter Four, we think that many Mexican women talked to us partly as a political calculation. Contact with the powerful is an opportunity to manipulate them and to enumerate the problems with local teachers, the need for electricity, clean water and jobs. On a more personal level, the opportunity for a woman to present herself is an opportunity to be the suffering, enduring mother who is the ideal Mexican woman. At both levels, we knew that we were being manipulated – but how often did we not know? There are many levels to this manipulation, many strands, many contradictions which we do not feel competent to deconstruct. We were too much the friendly strangers, too little the long-term participant observers. But we are confident that women worked hard, some or all of the time, to project an image of themselves and their communities which might, when published, possibly be of benefit to them. For several, telling the world about their troubles was an important part of telling their stories. Yes, we were able to exploit this attitude, we still have the power, the privilege and, in this book, the authority. But the women also changed all of us who recorded their life stories. By working on us, as people, they have reworked our views of the world.

These life stories sit rather oddly in the life history mode because their production was so quick. (We shall recount our methods in Chapter Four p. 54–7.) We agree with the Personal Narratives Group (1989: 261), 'When talking about their lives, people lie sometimes, forget a lot, exaggerate, become confused, and get things wrong. Yet they *are* revealing truths.' Many pioneer women want their stories told, and we hope to tell some here.

NOTE

[1] Throughout this book we shall use . . . to indicate a pause, and [. . .] to indicate where we have cut out a part of the quotation.

2

WOMEN PIONEERS IN THE TROPICS

WHITE WOMEN PIONEERS IN COOLER LANDS

> I took my wife out of a pretty house,
> I took my wife out of a pleasant place,
> I stripped my wife of comfortable things,
> I drove my wife to wander with the wind.
>
> Stephen Vincent Benét, 'Tom Brown's Body'

Isaiah Bowman, in his classic study *The Pioneer Fringe* (1931: 11), selected this verse to depict women's experiences as pioneers in the Mid-West of the United States. However, it is not women, but white women, that he celebrates. No indigenous or African-American women are mentioned in his account of the Mid-West, and in his chapter on South America only 'men and their sons' figure as pioneers. His account is not only an etic view, an outsider's analysis, but very much that of a white, male, North American outsider of his time. This chapter will show that in the literature on land settlement worldwide, in novels as well as academic studies, white women play their part while other women are almost absent. Bowman (1931: 12) quotes Hoover (1930), unveiling a statue of the pioneer woman at Ponca City, Oklahoma, 'The pioneer woman has played her part in the conquest of nature through all the ages.' But it is white pioneer women, the women among the conquerors, whose achievements are celebrated:

> What pioneering does to family life and education is largely written in terms of what it does to women. Probably the change that has come over the humans who seek life at the frontier, especially in their attitude toward help from the government for roads and schools, is largely due to the unwillingness of the women to stand the hardships and primitive life of the untamed land beyond the settled communities [. . .] Nowadays, and to an increasing extent, women are responsible for the flow of culture into the pioneering lands of the world; that is the case at least with white women and lands held by English-speaking sovereignties.
>
> (Bowman 1931: 12)

18

'At times we have the treacherous thought, Is it worth while? But glancing at our homes, here in the wilderness, there comes the thrill of achievement; the knowledge that where we white women are, with our children, civilization and higher moral standards must necessarily follow.'

(quoted by Bowman 1931, from Muriel Drayton 1931,
'The farmer's wife – in Kenya')

Our images of women pioneers come above all from tales of white women in the USA, where hardship on the 'frontier' of the West led to prosperity for many. It is important to examine this image, although the reality in poorer countries has been very different. Influential male historians, such as Frederick Jackson Turner (1921), gave very masculine accounts of the frontier and virtually ignored pioneer women. Despite this women pioneers are now widely studied by historians (Jensen and Miller 1980; Armitage and Jameson 1987; Jeffery 1979; Schlissel *et al.* 1988) and in geography and landscape studies, and the importance of class and age is becoming recognised (Norwood and Monk 1987). Increasingly, more and more evidence is drawn from the 'insiders' view', as emic accounts and private documents such as diaries and letters change the perceptions of 'how the West was won' (an outlook which we had garnered from books, newspapers and official records, coloured by the film industry). Unfortunately, our view is still through the eyes of one group of white women, as literate pioneers usually came from the north-east USA. Despite this, the documents are still important. They illustrate women's gains as well as their sufferings and some argue that the benefits from homesteading outweighed the penalties, that these were women of high status, partners in a joint venture whose relationships with men improved and whose opportunities widened (Kohl 1976; Harris 1983). Some of the sources maintain that women's independence and equality were limited to the pioneer days before about 1880 and that they later resumed their roles as household 'angels' (Frick 1982), while others stress the hardships (Schlissel 1982). The pioneer worlds were complex, with published works speaking of female submissiveness, purity and leisured domesticity, although women's informal talk and letters frequently did not (Jameson 1987). In contrast to the situations of most white women, indigenous and black women faced forced acculturation and discrimination (Jensen and Miller 1980; Reese 1991), perhaps especially because white men had relied much on indigenous women's skills, so that white women were jealous (Van Kirk 1987).

Until recently, it was novels which gave society 'the pioneer woman'. Willa Cather, Hamlin Garland and Mari Sandoz depicted and favoured women of strong will, self-reliance, independent spirit, determination, endurance and physical and moral strength (Yuvajita 1986), yet Mari Sandoz and Hamlin Garland also portray first-generation women through the eyes of their children, as victims of emotional and physical abuse by their husbands (Maples

1989). There are some fifty novels and short works by women about the experience of women pioneers on the prairies (Fairbanks 1983).

Recent writers have challenged the academic and fictional images of frontier women. Susan Armitage (1987) identifies three main stereotypes: the refined lady, too genteel for the frontier; the helpmate, who becomes a work-worn superwoman; and the bad woman of glamour and power who comes to a bad end. Armitage develops another image of the reluctant pioneer, desperately lonely, who has left behind her emotional supports, her family and friends, and suffers hard work and isolation. All these images would be easily recognised by pioneer women in Latin America, as would the evaluation of men primarily in terms of economic achievement, women in terms of the domestic sphere (Kohl 1976). They would recognise, too, the 'Cult of True Womanhood', an ideal from Eastern cities which confined men and women to separate spheres (Welter 1966). Women were to occupy themselves with the moral and practical issues around the home and the care of children, and to cultivate purity, piety, domesticity and submissiveness for the sake of husbands, fathers and children (Schlissel et al. 1988) as theirs was the higher spiritual nature. Men were to restrict their actions largely to the public world of production and business and to be aggressive and competitive. At least the role of helpmate was compatible with the farm economy for, as their husbands' helpers, women could work outside their proper sphere and still conform to the ideal. Latin Americans would recognise this ideal, too, as they would Schlissel's comment (1982: 155), 'If any idea joined the women to their men, if any expectation made the strenuous journey bearable, it was the idea that the move would bring them and their children a better life.'

If we study the work of men and women pioneers and the relations between them, we find that there were great variations over space and time even within the North American 'West'. Women could be homesteaders in their own right (Patterson-Black 1976), many widows owned farms and it may be that black, immigrant and poor women worked far more on the land than was admitted (Sachs 1983). Broadly, the image is that men ploughed, planted, harvested and cared for sheep, horses, cattle and pigs while women grew vegetables, canned them, made the butter, cared for the poultry, cooked (on coal or wood stoves), made bread, did the laundry, made clothes, mended them, ironed them, cared for the sick, did the housework, brought up the children and often chopped the wood, carried the water and made the candles and soap. Women's cream, butter and eggs were an important source of personal and perhaps family income. Women also dispensed hospitality and assisted in childbirth. Family survival, however, depended on flexibility and interdependence in work roles (Jameson 1987; Kohl 1988) and men could cook and even deliver babies while women could plough and would run the farm in the men's absence. In north-east Colorado 10 per cent of those acquiring homesteads were women, and many girls not only learned household duties but worked in the fields (Harris 1987) so that they grew up with more skills.

In identifying some points so familiar from Latin America, it is important not to overstate the similarities. For instance, most North American women seem to have been much less isolated. True, women's pain in moving west was expressed primarily in terms of leaving family and women friends, but nearly half of emigrant families travelled in groups with kin (Jameson 1987). Julie Roy Jeffery (1979) was struck by pioneer women's commitment to making new friends, visiting and building a new network of women. Even in ranch communities (in Saskatchewan at least) there was an active social life (Kohl 1976). Katherine Harris (1987) describes these networks as weak, but they counted for a great deal in women's lives and, above all, were socially sanctioned as they would rarely be for pioneer women in Latin America.

LAND SETTLEMENT

Pioneering in temperate lands, in North America, Siberia, Australia, New Zealand and to a lesser degree the southern cone of South America, carries images of hardship endured for subsequent prosperity, and of the ultimate achievement, very widely, of reasonable standards of living. On a much smaller scale, the same is true of Israel, which now seeks to teach poor countries how to repeat its own achievements. In medieval Europe, success often took generations. Harald Uhlig (1984: 90) illustrates the privations with an old proverb about clearing land in the Alps:

> To the father, death,
> to the son, still want,
> and only to the grandson, bread.

(No women appear in this saga!)

In the tropics, colonial settlement by whites using indigenous labour, as in the 'White Highlands' of Kenya, was often profitable – to the whites. Postcolonial colonisation of 'new' lands in the tropics by people from the same country, whether in Amazonia, Nigeria or Indonesia, has been another story. The preference for land settlement rather than improvements in existing villages is described by Ester Boserup (1970) as a colonial legacy, best discarded, and we agree with her.

A very few pioneers achieve prosperity, an example being Justo, who entered the Colombian rainforest in 1951 with Etilda, his two small boys and 'a machete borrowed from my brother', and thirty-three years later had title to 299 ha. of land and thirty head of cattle (Townsend and Wilson de Acosta 1987). Many fail altogether, many children die and many of the survivors live in great poverty. Justo, who built prosperity from destitution by hard work, good management and good luck is very exceptional. As Jeffrey Jones writes (1990: 135): 'Land settlement is not an activity for the poor [. . .] the final effects [. . .] have a tendency to favour the wealthy.'

21

Why land settlement?

There are six main reasons usually given for land settlement schemes.

1 To use 'new' land, as in the rainforests of Amazonia or the Outer Islands of Indonesia. Many of these 'new' lands are in the humid tropics, and it is with these that we are primarily concerned, for settlers in the tropics often suffer much for little while causing great damage both to the original inhabitants and to natural resources. Usually, the people who profit are city speculators, as in Brazil. Political considerations may be important. In Brazil, the TransAmazon Highway was built and Calha Norte is now being 'developed' partly so that no other country will occupy the land, while in Sri Lanka, the Mahaweli Scheme has been used to move millions of people belonging to the Sinhalese majority into potentially rebel Tamil areas. People in 'new' lands need a whole infrastructure and very often planned schemes require poor settlers to depend on seeds, credit, advice, even roads, which, through failures of administration, are simply not provided in time to be of value.

2 To invest in increased production, using dams and irrigation to intensify land use. This was an important colonial strategy, but many recent schemes have been much less successful. Again, all depends on administration and the new settlements are often desperately short of food, training, credit, infrastructure and services.

3 To resettle people displaced by big dams or new airports, towns, opencast mines or forest reserves. In practice, most displaced people are offered no land or jobs, and administration fails even those who are resettled. The classic example is the Volta Dam in Ghana where the displaced were often destitute, there was much environmental damage and the returns were poor. There are fears that the Narmada Valley Dams in India will be as bad and the Three Gorges Dam in China is even more alarming. Dams started between 1979 and 1985 promise to displace between 1.4 and 2.1 million people annually (Cernea 1993b). Michael Cernea (1991a) estimates that projects with World Bank financing in the 1980s caused the displacement of 1.6 to 1.8 million people and reports that impoverishment is a common outcome.

4 To settle nomads and shifting cultivators, which may appeal to governments who prefer sedentary citizens but not so strongly to those who are forced to settle.

5 To move people from unsafe sites after natural disasters, such as earthquakes, floods or drought. This often appeals to bureaucrats rather than the option of supporting people to rebuild and make safer their lives in their original homes.

6 To settle refugees. These new settlers have little choice, but their plight is desperate (Cernea 1993b). Some schemes, as in Tanzania (Armstrong 1987), have been relatively successful in alleviating suffering.

Planned land settlement is technical and social engineering and settlers feel at the mercy of planners and bureaucrats. Land settlement is often seen as a way of creating new, more equal societies on 'new' lands, but, whether in the Philippines or Peru, the settlers take their society with them and at best recreate it, often causing greater inequality than before. Academics and politicians continue to debate whether this inequality necessarily follows, or whether different methods could give better results.

David Hulme (1987) argues conclusively that the continuing popularity of land settlement (with governments and aid agencies) despite high expense and frequent failure can be understood only from an examination of the four main groups who influence the decisions. Politicians gain a supreme rhetorical device, a panacea with high visibility which can divert attention from other intractable problems such as a need for land reform. Land reform often threatens the powerful, and land settlement can be made to seem an alternative which threatens only the powerless, the indigenous occupants of 'empty' lands. Bureaucrats gain jobs and power in a highly appealing form. International aid agencies gain quick, large, 'off-the-shelf' schemes. Commercial interests gain contracts and profits. Together, these interests explain the continuing national and international failure to learn from experience or evaluations and tell us *why land settlement.*

Success in land settlement?

Of all strategies for rural development, land settlement schemes were always 'particularly failure-prone' (Hirschmann 1963), and 'very few programmes have achieved their stated objectives' (Oberai 1986: 58). Planned schemes in 1985 cost US$5,000 to US$10,000 per settler household (Scudder 1991). A.S. Oberai, in a major review for the International Labour Office (1988), does think that minor improvements could be made to land settlement through better planning, management and administration, but he lists an array of failures and problems. For him, land settlement is at best a palliative and other options should always be evaluated: 'Looked at in purely economic terms, few land settlement schemes are likely to be "viable"' (Oberai 1988: 27). Thayer Scudder, reviewing projects for USAID (1981) and the World Bank (1985, 1991), is also critical of past schemes, but he is much more positive about the possibilities – if only a more social scientific approach were employed. Perhaps he recognises that settlement will continue to be popular and seeks to make it less disastrous. He professes great faith in the potential of good planning, of technocracy, although hardly from positive experience: 'The majority of government sponsored settlement schemes cannot be considered a success in terms of either direct or indirect benefits' (Scudder 1981: 357). His opinions are important, as the World Bank is now the major international donor in land settlement (Scudder 1991).

Many academics have tried to develop guidelines for land settlement.

23

W. Arthur Lewis sought to do so in 1954, but noted even then the contradictions between the popularity and the expense and problems of land settlement. For him, there are seven factors upon which success depends:

1 Choosing the right place;
2 Choosing the right settlers, with some capital of their own (all settlers being 'he'!);
3 Supporting spontaneous settlement, but, if planned settlement is necessary, making enough investment;
4 Supplying settlers with minimum capital;
5 Using compulsion as appropriate;
6 Awarding enough land to make a living;
7 Providing security of tenure.

Hundreds of studies over the last forty years have looked at land settlements and made recommendations (often the same as Lewis's!), but settlement is still expensive and problematic. A remarkable industry flourishes in feasibility studies, environmental impact studies and evaluations of land settlement. There have been thousands of publications, to little effect. For Gary Palmer (1974) this industry is ecological imperialism, with the unfortunate settler struggling in one niche in the new natural environment and in another niche in the bureaucratic environment of the national and international system. Many settlers certainly feel exactly like this, and say so.

The literature is highly imperialistic and most studies are based on questionnaires designed by outsiders, outside the project and often foreign. The views of outsiders are always decisive. W. Arthur Lewis (1954) and most work since take a highly colonial and technocratic point of view (even though W. Arthur Lewis was from St Lucia), empowering the bureaucracy to control and to decide for the supposedly ignorant settlers, but having little positive impact, for W. Arthur Lewis' recommendations have had to be echoed by A.S. Oberai (1986) and the World Bank (1985)!

Criteria for 'success'

One colonial power which did pay attention to the characteristics of settlers was the Dutch who had substantial experience of land settlement at home. From 1932 to 1941, when selecting settlers to be moved from Java to the Outer Islands of the Dutch East Indies, they paid close attention to wife and family as well as to farming experience, the best option being always to move a whole community (Pelzer 1945). Although 'top-down', this process did take insiders, settlers, into account. But this colonial attention to *social considerations* disappeared in postwar economics.

Several writers see the *quality of administrators* as the crucial feature, in Africa (Chambers 1969), Latin America (Dozier 1969) and Indonesia (Arndt 1988). Such a range of problems is involved that the planning of land settlement is extremely demanding in administrative terms: 'The settlement process is a very complex one and few governments understand or can cope with this complexity' (MacAndrews 1979: 127). Good administration can be extremely costly. That may explain why many writers advocate either planned schemes with high expenditure or 'spontaneous', low-cost colonisation where the state provides a little infrastructure and settlers make their own arrangements, for settlers are seen as being more efficient than bureaucrats unless expenditure is very high indeed. In the humid tropics, some 75 per cent of land settlement is spontaneous. Spontaneous or 'lightly administered' colonisation has been advocated as the most cost-effective for Latin America in a highly influential study (Nelson 1973), for Bolivia (Zeballos-Hurtado 1975), for Africa and the Near East (Higgs 1978; Belshaw 1984), for the Philippines (James 1979), for Indonesia (Guinness 1977; Sumarjatiningsih 1985; Vayda 1987) and in general (Beenstock 1980). Thayer Scudder (1985, 1991) proposes a careful mix of planned and spontaneous settlement. For similar reasons, some advocate privately financed schemes (James 1983), which have had some popularity in Brazil at different dates (Katzman 1978; Butler 1985).

Some authors have paid attention to *insiders*. Although technical concerns have dominated the literature, Thayer Scudder called in 1969 for attention to socio-cultural issues among the settlers as did Raymond Apthorpe (1968). Writers from Israel also have this focus (Weitz 1971; Maos 1984; Levi and Naveh 1989; Levi 1989) and, again, many see the problems as soluble and have specific proposals. The Israeli school, however, is largely blind to gender, as are many other writers calling attention to social issues (van Raay and Hilhorst 1981). The early exceptions were Colson (1960), Scudder (1969, 1981, 1985, 1991), Moris (1969) and Chambers (1969). For Scudder (1985: 127), 'Although there are many reasons that most new land settlements do not live up to planning expectations, inadequate *attention to settler families* and the communities in which they live is certainly a major explanation' (emphasis added). Robert Chambers and Jon Moris (1973) called attention to insiders, showing that evaluations by planners, managers and settlers necessarily differ, so that they cannot be combined into a general evaluation: the etic 'facts for the outsider' differ from the emic 'facts for the insider'. The World Bank's 1990 directives on forced resettlement indeed require not only more consideration of the people displaced but consultation with them (Cernea 1991a, 1993a), and, as a result, the Brazilian Power Sector has, at least in theory, proposed new policies (Fernandes Serra 1993) and Mexico's national power company has made new efforts (Guggenheim 1993).

We have just seen that the emic and the etic views cannot be combined into a single general evaluation. Other generalisations about land settlement are also problematic. One example is the frequent argument that 'spontaneous' land settlement is cheap, for others argue that the success of the scheme is proportionate to the cost (Gosling and Abdullah 1979) and that spontaneous settlement has produced and will continue to produce extensive environmental damage (Sandner 1982). The explanation for the different outcomes may lie in the schemes chosen for analysis (Hulme 1988)! The thousands of studies limit the completeness of overviews, and adjoining areas may yield contradictory experiences (Uhlig 1984).

Attention to insiders was largely absent, except in Indonesia, for the first twenty years after the Second World War and it is still far from universal. In the last twenty years there have been three reactions to the top-down approach as more attention is paid to the repercussions for insiders, for indigenous populations and for the environment. It has been recognised that even land settlement which, as in Malaysia, has made the settlers prosperous, may not be socially successful and may both destroy the previous inhabitants and be ecologically devastating. The criteria for 'success' have become even more debatable and much interest has developed in alternative forms of change (for example Panama: Joly 1989; Ecuador: Uquillas 1989; Colombia: Bunyard 1990).

Sustainable livelihoods?

In proportion to the scale of poverty in poor countries, land settlement is a minor issue for, in the developing world as a whole, the expansion of the farmed area is a mere 0.1 per cent each year, and 75 per cent of this is spontaneous. Planned land settlement only provides for a small proportion of the annual growth of rural populations, let alone the existing population in the source area. Yet there have been massive regional impacts on local people and on the environment. In Indonesia, for example, over 4 million people have moved over the years, mainly through planned schemes, and in Thailand 5 or 6 million, mainly spontaneously (Uhlig 1984). In Latin America, some 25 million people now live in areas of recent settlement and most migration has been spontaneous. The state, however, is always involved, at least in granting title to land and in the provision of infrastructure, particularly roads. Very few spontaneous colonists will go where there is neither access nor hope of access to markets.

Many of the world's 200 million indigenous or tribal peoples have suffered under this occupation, often being displaced by agrarian systems much less sustainable than their own. Good feasibility studies are rarely performed before a scheme is developed. Spontaneous migrants may have no experience of the environment and a planned scheme may incorporate an inappropriate farming system. In the early 1970s, the Mexican government planned to conquer the rainforests of Uxpanapa (in the Isthmus, p. 67) using the latest technology of improved rice, pesticides, bulldozers and combine harvesters. The pests proved uncontrollable due to the high rainfall and the mud defeated the combines so that the unfortunate Indians who had been forced to migrate to the scheme had to fall back on their traditional techniques to survive (Ewell and Poleman 1980). There have been further experiments in Uxpanapa with rubber, cattle, teak and oranges, but all have failed. Contrary to indigenous systems of production, many of which are extremely diverse and highly sustainable, introduced systems frequently are of monocultures, often of annual crops, which are at great risk from pests or from variations in the weather or the market (Townsend 1985). Usually, livelihoods for the settlers could have been provided more cheaply and with far less devastation in areas of the country already settled. Although we are more aware of the sufferings of indigenous peoples now that they have learned to organise and to use bodies such as Survival International to publicise their plight, this is not saving them.

Logging and deforestation for land settlement have produced a sustained attack on the world's rainforests and other ecosystems. Some of the mechanisms here are strange. In Brazilian Amazonia, no crop or livestock production is profitable unless the rising value of the land is taken into account, but land is one of the few investments which increase in price faster than inflation, and to clear forest for cattle ranching is an effective way to claim land. Land speculation has created large areas in which jobs are few and poorly paid, little is produced and the land is simply held for speculators.

This is not the place to discuss whether the loss of a species is a loss to humanity, whether tropical rainforests are critical in the global carbon cycle or whether cattle and rice paddies promote global warming by releasing additional methane. The local and regional destruction of soils, damage to rivers, loss of biodiversity and effect on climate (Salati 1987) are unquestionable. We do not have the knowledge to convert large areas of 'new' lands to agriculture and pasture on any sustainable basis. Deforestation is not the product of any ineluctable force, such as pressure of population or international capitalism, it is the outcome of specific national and regional developments, each with its own history and political economy.

Two essentials for the future of the tropical world will be the retention of large areas of forest and the development of better land uses in areas already occupied. For both, 'the achievement of long-term sustainability in the lowland tropics depends on the development of markets appropriate to that environment' (Jones 1990: 131). Forests will be saved only if we can recognise and utilise their economic value, which can be high. In eastern Peru, the value of the fruits, nuts, latex and other forest products from a hectare of tropical forest would exceed those from logging and ranching (Peters *et al.* 1989), if only the marketing structures existed for the sustainable products. The 'extractive reserves' developed in Brazil may contribute here, for all their problems (Hecht and Cockburn 1990). There is a great variety of crops which are ecologically appropriate and sustainable but which are little grown. If the products of forest and field could only be marketed, there would be immense possibilities for managed forests, agroforestry and other sustainable forms of production in the humid tropics.

Invisible women

Of the thousands of studies of land settlement in less wealthy countries, only a tiny proportion mention the women who constitute half the settlers, or pay any attention to gender. In Brazil, President Médici proclaimed in 1970, when announcing the building of the TransAmazon Highway, 'We shall move men without lands to lands without men.' In practice, of course, women and children are also moved, and usually there are people there already, for there are no 'lands without men'. All these invisible people may suffer appallingly. Projects are often designed in terms of 'the small nuclear family in which the man is head, breadwinner and owner of the allocated land' (Schrijvers 1988: 45). Planners appear to have an ethnocentric, Western idea of moving 'a man with a family', rather than reflecting local custom. Even spontaneous migration often involves a painful disruption of the extended family. As Robert Chambers observed in 1969, 'the lot of women is often worsened by settlement', through loss of their rights to land, reduction of their access to income and to income-generating opportunities (while their husbands gain a new monopoly over income), an increase in their workload in the fields and home and a loss of social life (Chambers 1969: 174–5). (Since Chambers' book was extremely influential, the continuing invisibility of women in land settlement is important as the whole culture of land settlement seems averse to their recognition.) Susie Jacobs (1989) identified the same problems, and a very similar list was reported to USAID in 1981 (and repeated in 1991) by Thayer Scudder, apparently independently and again with little visible impact. In 1985 he claimed that there were strong similarities between settlers: 'People who undergo land settlement, whether voluntarily or compulsorily, respond to the process in predictable ways . . . This is true across cultures, partly because the stress of relocation limits the range of coping responses of those involved' (Scudder 1985: 123). It is also partly because the planners of land settlement, commonly prefer to follow inappropriate international models rather than respond to local culture.

In 1981 Thayer Scudder assessed almost 100 schemes and found only two where the woman pioneer was considered at all in the selection of settlers. The exceptions were the Mahaweli scheme in Sri Lanka, where the only thing noted about the wife was whether she belonged to a women's organisation, and the FELDA schemes in West Malaysia, where both spouses were interviewed (scoring a maximum of 40 points – 27 for the man, 13 for the woman). Scudder (1969, 1981, 1985) argues that both spouses should always be interviewed, although they very rarely are.

Women's role in production in land settlement is extremely varied. In Nigeria, at the Ilora Farm Settlement (see p. 30), all women spend about a quarter of their time farming (Spiro 1987), which would be very unusual in Latin America. Also in Nigeria, on the Kano River Irrigation Project, Muslim women are paid for their farm work by their husbands and maintain independent budgets (Jackson 1985). In rubber-growing settlement schemes in Malaysia, 75 per cent of settlers' wives tap rubber, and in oil palm schemes, 56 per cent of settlers' wives harvest oil palm – but the wages are paid to the husbands (World Bank 1987). James Weil (1980) records that, among highland Indian settlers in lowland Bolivia, women do about a third of the agricultural work and men about a fifth of the work in the house (although very little child care). At Tucumá, a private project in Brazil, women often support the new farm through off-farm work, washing clothes, working in hotels, running small shops, selling garden produce, while the men take the first crop (Butler 1985).

Women's common problems

No title to land

When land is distributed or redistributed, in land settlement or in land reform, planners are concerned that holdings should not be subdivided and therefore prefer that the rights to the land are vested in one person (usually the male 'head of household') and are transferable to only one person (often the eldest son) (Hahn 1982; Palmer 1985; Deere 1987; Jacobs 1989). This occurs both in societies where few women have rights in land and in those where farm women normally have such rights, as in Sri Lanka (Schrijvers 1988), for uniformity has been imposed on an array of cultures. Therefore many women pioneers who grew up in communities where women had rights to land now find themselves landless. They suffer further instability as marital breakdown could threaten their livelihoods and those of their children.

Less income

In most rural areas, women have considerable sources of income, in the fields, as artisans, or in petty trading. Often it is their incomes which feed and clothe the children. As pioneers they may lose their rights to use the land, they may no longer have common land to pasture small livestock, their traditional raw materials may be unavailable and their access to markets may be poor. If the women have always had an independant income, they and their husbands may have no experience of pooling family income. Therefore, the women may now be forced to extract household needs from the men, who are not used to supplying them (Burkina Faso: Palmer 1985; Nigeria: Spiro 1985; Malaysia: World Bank 1987; Sri Lanka: Ulluwishewa 1989; Colombia: Townsend 1993b).

More work

The workload of pioneer women is often greatly increased, especially in the early stages. There is often much more mechanisation of men's work in ploughing and clearing land than of women's sowing, weeding and harvesting which come under increasing pressure. Occasionally, women's work is reduced: Robert Chambers (1969) cites the Gezira, where Arab tenants responded to enhanced incomes by withdrawing their women from the fields, and Upper Kitete, where mechanisation of the whole wheat cycle released women from several tasks. But he regards such cases as exceptional. Credit and extension services have generally been closed to women. Liam Pickett (1988) reviewed the role of co-operatives in settlement schemes around the world. Specific provision for women was found only in the Sudan, where housewives had themselves formed co-operatives, and in Indonesia, where he reports specific training for women colonists. He calls the general lack of provision a planning failure.

In their home areas, women's childcare, housework and agricultural and other tasks may have been shared with other women but now they are often isolated with husband and children, far from friends, relatives and any support services such as health facilities, schools, electricity and clean water. Thus women's 'double burden' of productive and domestic work is often expanded.

Social isolation

Robert Chambers (1969: 175) writes of the 'social nakedness that has to be endured by settlers'. This isolation compounds the tremendous stress of relocation emphasised by Thayer Scudder (1969, 1985, 1991). Men must make new social links for the very survival of the farm, but women may have no such tradition. In the very different societies of Colombia and Sri Lanka, pioneer women say that in their home villages they felt safe to move about freely, but now they live among strangers and there is no one with whom they can leave the children, so that their freedom of movement is severely curtailed (Lund 1981; Palmer 1985; Ulluwishewa 1989; Townsend 1993a). In Colombia, Janet Townsend has interviewed women who have delivered their own babies, not because it is customary for them but because they 'knew no-one to ask for help' (Townsend and Wilson de Acosta 1987). In the spontaneous settlement of the Chonburi Hinterland in Thailand, regarded by Harald Uhlig (1984) as economically successful, families are painfully divided by their different descent and illegal status (Schentz and Leitzmann 1984).

Case-studies

The Ilora Farm Settlement in south-western Nigeria was designed, after a visit to Israel by the Minister of Agriculture, to keep the educated young in farming by creating a modern, co-operative farming community, but it suffered from a high rate of desertion (Roider 1970; Spiro 1985). The settlers had no role in managing the Settlement. Women were merely spouses. They were granted no land rights, nor rights of inheritance, and if a man died without male relatives, his wife had to leave the scheme, even if she had been doing all the work. The people in the Settlement are Yoruba, and it is customary for man and wife to live their economic lives almost independently, with little or no pooling of income. Traditionally, women's income is earned primarily through trading, although land is also allotted to them to farm. The Settlement allocated no land to women, and created no trading opportunities and, in practice deprived them of access to credit. Despite the commitment to modern technology on the land, there was no introduction of piped water or of labour-saving in the processing of food (Roider 1970; Spiro 1985). Roider (1970) reported that wives did not like the way of life.

In Kenya, many women found the Mwea Scheme in the 1960s 'an intolerable place to live', and some abandoned their husbands (Hanger and Moris 1973). The failure to recognise women's rights and needs is seen as one of the great weaknesses of Mwea. Fields were distant, water was polluted, fuel was scarce, prices were high and women no longer had access to land to meet their traditional responsibility of feeding the family. Instead, they had to work for their husbands without pay, and, for the first time, depend on their husbands for the necessities of the family. They had to learn to grow and to live on rice, which was still disliked as a food ten years later, while much of the profit from the rice was spent by the men on beer.

The Mahaweli River Development Scheme is the largest and most expensive development project in Sri Lanka, yet the settlers suffer chronic undernutrition. Joke Schrijvers (1988) attributes the poor nutrition to planning which cut women off from productive resources and particularly from the production of food. Sinhalese tradition is that all sons and daughters inherit, but special legislation for the Mahaweli project requires a single heir, who is usually a male. In the old villages, women do most of the work on unirrigated fields, look after dairy cows and cultivate kitchen gardens. Men do most of the work on the irrigated rice, but women can feed the family with limited support from their husbands. In the Mahaweli project, families have an allotment of irrigated land and a garden plot, but often no unirrigated land. There is no common land to graze cattle and often no water for the tiny garden plot. As at Mwea, men control the irrigated land, the decision-making and the products (Ulluwishewa 1989) and women must work for them. Training is of men by men. In the old villages, much work is shared and bathing and the collection of firewood and water are group occasions for enjoyment. In the Mahaweli, women have minimal social interaction with other women, but must learn to live in a nuclear family in which they have lost much power. Personal relationships intensify between husband and wife, and the woman has little social contact outside the home (Lund 1981). Thayer Scudder (1991) notes that very little social science expertise was used by agencies funding the Mahaweli schemes.

In the Jengka Triangle Projects in Malaysia, the World Bank (1987) reports that an inconsistent application of Islamic principles has resulted in an erosion of women's rights overall as compared to the traditional *kampung*. The production of rubber is important, and farmers' wives form a substantial part of the labour force, yet technical advice is regularly provided only for the man, who is paid for the woman's work. Only married men may enter the schemes, the title is in the husband's name and, in the event of divorce or if the husband takes a second or third wife, the first wife usually has to leave the scheme (World Bank

31

1987). Land settlement schemes in Western Malaysia are widely seen as among the developing world's few successful examples, but serious difficulties are emerging. As services are poor and there is little non-agricultural employment, young people are leaving for the towns (Bahrin *et al.* 1988).

In highland Bolivia, women control a farmyard of small livestock, own and rent land, use common land and earn a living by the sale of handicrafts and petty trading. When they move to farms in the eastern lowlands, they lose all these resources (Hamilton 1986). The new title is for the man. Further problems arise because the woman does not know how to build up a stock of pigs, sheep, goats or poultry in the warmth and humidity of the lowlands – if she has managed to bring any from her former home, they will probably die. Women cannot get the raw materials for their crafts or sell the finished products. There is no common land in the lowlands. In the San Julian scheme documented by Susan Hamilton (1986), only men are trained as health workers, although women are unwilling to be examined by men who also tend to be out in the fields when children have accidents.

In many new ranching areas of the Latin American tropics, it is difficult for rural women to generate income (Lisansky 1979; Hecht 1985; Townsend and Wilson de Acosta 1987). From Mexico to Argentina, almost everywhere that colonisation involves ranching, women not only lose control of the proceeds of their work but are excluded from production and reduced to housework and prostitution (Hecht 1985; Court 1986; Townsend 1993a). Even in the towns, the only work which will keep a woman-headed household is prostitution. Yet in the highlands women care for cattle, and in eastern Colombia they share in the care of the same breeds of cattle that in other new settlements are said to be too savage for them (Meertens 1988; see also p. 43).

There are few apparent examples of land settlement where women have succeeded. Mishamo, in Tanzania (Armstrong 1987) is a settlement for refugees where women's groups appear to be among the more successful, with communal productive activities ranging from farming to brewing, making clothes and craftwork, while day care centres for preschool children operate along every feeder road. At first, agricultural training was only for men, but, since women were doing most of the work, this training was ineffective and later each village had two agricultural field assistants, one man and one woman (Southey 1984).

In Zimbabwe between 1980 and 1985, more than 250,000 people were resettled from Community Areas on to what had been commercial farms owned by whites. Susie Jacobs (1989) describes this resettlement as relatively favourable to women although some familiar problems reappear. The new farms are held by the male 'household head' and although legally men or women could apply for land, no settler in her survey knew that. A divorced woman has no right to stay on resettlement land. Most farms depend on women's unpaid labour. Nevertheless, she concludes from comparative calculations that women's access to land is no less than in the Community Areas and possibly more. Because these were large, commercial farms, there is often no water supply or clinic, school, shop or market and transport is usually poor. Therefore, the women's workload is increased, but so is men's, and the loads are perhaps more equal. Women are more isolated than men, and families are more isolated than before, so that the husband's power may increase and he may make more intensive use of his wife's labour or husband and wife may become more mutually dependent. The nuclear family is 'double-edged', for women have more say in the family and more influence over their husband but more dependence on them. Women do not feel that they have gained in autonomy with the resettlement, and they complain of male violence and personal and sexual problems. Yet most women feel that they have benefited financially and many say that their husbands behave better in the settlements, work harder, spend more on the family, drink less, are less violent and respect their wives more. Thus, not all land settlement makes things worse for women.

CONCLUSION

In rich countries, we have positive images of land settlement from our own history and literature. In poorer countries, land settlement has been and is a strategy for rural development that is popular with governments and aid agencies. Yet planned settlement rarely appears to be economically successful in proportion to its cost. It is commonly an expensive option in comparison to investment in settled areas, but remains popular with governments (for political reasons) and with powerful interest groups. Most successes are either where it has been possible to imitate local plantations, as in Malaysia, or where it has been possible to extend local village agriculture on to new terrain, as in Sri Lanka before the Mahaweli scheme (Farmer 1957). Otherwise, few schemes meet their targets. Spontaneous settlement is cheaper, but is often simply an extension of rural poverty with success for only a few, as in the Chonburi Hinterland, Thailand (Uhlig 1984) and in Amazonia (Hecht and Cockburn 1990). Spontaneous settlement is not an activity for the poor, and the final effects tend to favour the wealthy (Jones 1990). Social aspects have too often been ignored, as have sustainability and the fate of the existing inhabitants.

In North America, women pioneers were at first ignored by historians, but were later 'discovered' because of their diaries and letters. Women pioneers from poorer countries would recognise many of the images. Only a small number of studies of land settlement in poor countries take account of women or gender. Oddly, many of these studies report similar findings, as women pioneers tend to lose rights to land, access to income and social life while facing an increased workload under great personal stress. The scientific, international, outsiders' view tends to be based on questionnaires answered by men and to ignore women. When it does notice them it records their disadvantage, but to little effect.

3

IN THE COLOMBIAN
RAINFORESTS

COLONISATION IN COLOMBIA

As we saw in the previous chapter, land settlement is a popular, expensive and usually unsuccessful strategy for rural development. A massive literature has been produced (most of it ignoring women) by outsiders, experts who seek to be objective, rational and detached in order to establish the 'truth'. This chapter will describe our field research with women pioneers in Colombia. This, too, was research by outsiders, oriented towards policy. We sought to build up evidence, to be 'objective', but at least to examine women's roles. We shall see the limitations of this approach.

Colombia is a prime example of the power of myths in land settlement. For decades, the nation saw land settlement as an escape, a solution for old rural areas where the good land was monopolised by a few. In the 'new' lands, a new democracy of family farms was to be built. Today, these 'new' lands are the core of the criminal and revolutionary violence which plagues the people and threatens the state. For all Colombians, everyday life is coloured by violence. People talk about murder in Colombia as they do about the weather in Britain, and for men between 15 and 45 murder is the commonest cause of death. Many times since 1945, Colombia has had the highest rate of death by violence in the world for a country not at war. Over this period, however, the map of fear has been transformed. In the 1930s and 1940s, intellectuals saw the frontier of settlement as the place of democracy, of freedom and of the family farm. In the 1950s, tens of thousands of peasants fled from the bloodshed of the long-settled highlands for a new life in the tropical lowlands, and the government itself saw planned settlement as a solution to problems of rural poverty in the Andean core. Since the 1970s, however, pioneer regions have been the most bloody rural areas.

The great mass of the Colombian population still lives in the cooler highlands (Map 2, p. viii), but the last forty years have seen the occupation of the tropical valleys and some of the eastern plains. Colombians saw great hope in this pioneering. In the nineteenth century, pioneers had settled on the warmer slopes of the western chain of the Andes, in Antioquía, which became, in the

34

early twentieth century, the most prosperous area of coffee production. The Colombian Federation of Coffee Growers presented this experience as the rise of family farms and democratic society (LeGrand 1989). This interpretation was celebrated by a geographer, James Parsons, in his classic *Antioqueño Colonization in Western Colombia* (1947), in sharp contrast to the great estates and impoverished peasants of the eastern Andes. For Paul Oquist (1980), the frontier was a real escape valve in the nineteenth and early twentieth centuries, and it was the closing of that frontier, when all the land in the Andes was taken, that gave rise to the first wave of major agrarian conflicts from 1925 to 1935. For the government, the frontier seemed a real alternative to agrarian reform, and under Law 200 of 1936, anyone who cleared and settled publicly owned land became entitled to own it.

In 1948 the Violence, which lasted for a decade, broke out in the interior. Conflict centred on land and in some regions half the farmers were driven out, while hundreds of thousands of people suffered terror, bereavement and scorched-earth policies (Sanchez 1992). Because one focal area of violence was the region of Antioqueño colonisation (LeGrand 1989), scholars began to rethink and to show that the 'democracy' of this colonisation was a myth: it had been manipulated by entrepreneurs bent on accumulating capital and land (Palacios 1983). As Catherine LeGrand (1986) has demonstrated, the history of land settlement in Colombia was a struggle for the land and labour between pioneers seeking to make farms, on the one hand, and merchants and landlords, on the other. This struggle continues into the 1990s.

Migration to the lowlands, both planned and spontaneous, continued after the Violence. In the 1960s, the pattern was still of 'homesteading' first, with pioneers trying to establish small farms, then concentration of the land into ranches and commercial farms (Townsend 1976). It was profitable to grow crops in the ashes of the forest, but as the plant nutrients were leached out of the ashes and the weeds began to grow, it often did not pay to buy the commercial inputs to continue cropping. As in Brazil (Hecht and Cockburn 1990), prices and credit were and are much better for meat production and for large farmers, so that cattle are a much less risky investment (Townsend 1985). Thus, settlers would try to make enough profit from the newly cleared land to buy cattle, and usually went bankrupt and sold to ranchers if they could not. In the 1970s, many pioneers found that there was one crop that was profitable, one crop that would enable them to keep their land and make a viable farm: marijuana. Later, the processing and even production of coca for cocaine became important in frontier regions, where airstrips, laboratories and fields could be relatively well hidden. The population of coca-producing frontier regions doubled between 1978 and 1985 (Molano 1988). Illicit coca is Colombia's leading export and the drug mafia is a small but supremely powerful group, both nationalist and anti-communist. In pioneer areas, some guerrilla movements organise local government; some levy a tax on coca, most levy one on cattle; some also practise kidnapping. Right-wing paramilitary groups have developed all over the country. Over the last ten years, successive governments have tried to come to some agreement with the guerrillas, seeking perhaps to isolate the drug traffickers. Everywhere, there are professional killers who kill for a fee, as there were in the Violence forty years ago.

It is evident that in Colombia, neither planned nor spontaneous land settlement has resolved the problems of rural poverty. On the contrary, it has been the focus of conflict. Violence in Colombia in the 1990s is much more complex than in the 1950s: 'Clashes between the army and guerrilla groups are frequent; drug traffickers threaten all who oppose them; and more than one hundred rightist paramilitary bands murder with impunity' (LeGrand 1989: 5).

THE MAGDALENA MEDIO

The Magdalena Medio is the floor of the middle Magdalena valley, between the central and eastern chains of the Andes, hot, wet and, until the 1960s, rainforest. Most of the rainforest has now been cleared and ranches occupy vast areas, carrying perhaps one cow per hectare where the pioneers once had intensive plots of plantains, maize and beans. Angel, working as a labourer at La Payoa (see pp. 39–47), had been one of the early pioneers. He told us in 1987: 'The poor never escape from poverty: we who opened the forest have nothing.' Guerrillas, drug traffickers and paramilitary death squads have complicated the underlying conflict between pioneers and ranchers until it has escalated into a dirty war that has subjected the entire civilian population to constant terror (Rementería 1986). The military continue to be involved in terror and torture on behalf of the landlords and have helped to set up paramilitary groups.

People have continued to struggle here for land and livelihood in the face of almost constant fear. In the 1960s, pioneers feared the guerrillas, who might take, although they did pay for, their stored food, but they feared the army much more, for the army, they said, would first accuse them of supporting the guerrillas and then rape, burn and kill. Both were described as dangers almost incidental to the daily hazards of making a farm from the rainforests. As outsiders, we were safer than they, for the army gave us military permits, although they stopped and checked us many times and ambushed us once. The guerrillas we did not meet. We had been advised by geologists exploring for oil that, if we did, we should hand over all food and medical supplies, treat any injuries and all would be well. We carried no weapons (with guns we would have been worth attacking) and camped each night by someone's house, asking permission to sling jungle hammocks from the trees and to leave cameras in the house, in the dry. People were usually scared at first, but there were never any difficulties for we were privileged as outside the conflicts.

Today, fear of the army seems less acute, but anyone with thirty head or more of cattle is said to be paying tax to the guerrillas for fear of kidnapping. Paramilitary death squads such as MAS (the Anti Kidnapping Movement) are referred to with sick terror. Everyone has lost a friend or relative, often not knowing whether to the army, the guerrillas, the drug barons or the death squads. Safety is such a problem that we were only able to undertake a study because we were commissioned to do so by Acción Campesina Colombiana (a splinter group of the national peasant movement) and the Hydrological Section of the Ministry of Agriculture. Are things better or worse than twenty-five years ago? Incomes are higher, fewer children are severely malnourished, services are better – but fear is still commonplace.

In the 1980s, we carried out survey research in three contrasting areas of the Magdalena Medio to document women's roles in colonisation: one remote, frontier area, the Serranía de San Lucas, one high-tech land reform scheme, El Distrito, and one idealistic cooperative, La Comuna Integral de Payoa (Map 2, p. viii).We used a questionnaire survey and informal interviews, imposing our own agenda in search of an orthodox etic account. We had some considerable surprises.

The Serranía de San Lucas

The Serranía forms the northernmost tip of the central range of the Andes in Colombia and is accessible only by river. It has been intermittently held by the FARC and ELN guerrilla groups, but in 1984, when we conducted our interviews, there was a national truce and the area was peaceful. The conditions people were prepared to tolerate show how desperate they are for land. The settlers knew that the bloodshed might begin again at any moment, and the area is hardly attractive to farmers. It is hilly and extremely wet, soils on the granites are generally poor, there are no roads and, as almost all transport is by mule and boat, it is expensive to take goods to market and difficult to make a living. (It costs many times more to move goods by mule or boat as by truck.) Pioneers must build huts in the forest from the sticks and leaves to hand, and must live, cook and wash in clouds of disease-bearing mosquitoes and sandflies, without electricity or running water, far from a shop, a teacher or a doctor. Health problems are pervasive. People relieve themselves near the house and catch hookworm through their feet from other people's excreta. Everyone has worms. Women die in childbirth, men under falling trees, children of tetanus, diarrhoea, respiratory and childhood infections. Few have enough to eat. Yet they are fighting to make farms.

Colonisation in San Lucas began in the 1950s but was 'bottled up' because no roads were built. As a result, although most of the forest has been divided into farms and much of it titled, some forest still remains to be cleared and burned, and few ranches have appeared. Small farms still survive and few people are landless, although in the dry season many men have to work on other farms for a wage. San Lucas supplies a small amount of cheap food to the towns, and is an important source of low-cost labour, for men migrate seasonally to agricultural areas, women and girls permanently to the towns. Farmers produce much of their own food but above all grow maize for sale in order to pay their labourers and try to make enough profit to buy cattle. Human labour is the principal means of production and technology remains very simple. The main tool is the machete and the main form of transport to the river is the mule. Under these conditions, families have many children and households include other relatives and hired labourers. Living conditions are poor and services few and far between. The real profits are in commerce and ranching and everyone wants to participate in these. The example of Alicia, Justo's daughter (see p. 21), is typical. Her husband has 35 ha. but can only afford to crop 5 ha. He grows their food, buying only lard, sugar, soap, coffee and cooking oil. They have no cattle, but rent out pasture to pay for fencing wire, and hope to get a loan for cattle in the future. For others, illegal activities have an attraction. Some farmers had built their prosperity on past marijuana production, and the small airstrip by a gold mine had been closed because of suspected cocaine trafficking.

In 1984, we set out to study child nutrition and gender divisions of labour in San Lucas. We interviewed a random sample of one household in ten in the village of Papayal, and every household along the forest trail from La Pacha to Norosí, using a questionnaire and an informal interview, and weighing every child under the age of 5 (Townsend and Wilson de Acosta 1987). For us, there was much beauty and people were kind and friendly. But we had been vaccinated, we boiled all the water we drank, we knew about hygiene and we were well-fed: our health and education came along with us. We did not share their privations or experience their lives.

We shall reserve our findings on the gender division of labour until later in this chapter, when we compare San Lucas with other pioneer communities (pp. 40–2). San Lucas is the only pioneer area where we studied child nutrition, and, therefore, our focus here will be on nutrition. Families are large, for people in San Lucas need to produce all the labour they can. Of the 114 women in the survey, a third were pregnant or lactating and twenty had had more than ten pregnancies: 'the boys are our labourers' and the girls can migrate to the cities. Despite the large families, the first objective of farms is investment in the future rather than self-provisioning and perhaps as a result nearly half the small children are underfed. A few eat earth, probably in search of missing minerals, as they have long done in the Magdalena Medio (Townsend 1976). In the 1960s, the cattle in the Magdalena Medio were so mineral deficient that they readily broke their legs (Townsend 1976). In San Lucas now, mineral supplements, vitamins and inoculation are seen as essential for cattle, but not for most children. In the households we studied, one child in six aged under 5 was severely malnourished (being less than 80 per cent of the expected weight for their age). We found no sex difference in malnutrition or child survival. Hunger is not a direct measure of poverty, however, for there were malnourished children in households with over 100 ha. of land. It is perhaps a measure of priorities, for none of the severely malnourished children had a head of household as a parent (Townsend and Wilson de Acosta 1987). Their parents might be resident workers, especially (women) cooks, or younger children of the owner. (No woman head of an independent household had young children.)

El Distrito

Across the river, only 100 km from the mules and machetes of the inaccessibile Serranía de San Lucas, lies a small high-tech land settlement scheme, El Distrito. El Distrito is another world, of fenced fields, gravel roads, aerial crop sprayers and combine harvesters, only five hours from the city of Bucaramanga by gravel road. Here, pioneers cleared the forests but ranches had taken over. In 1965 the state bought some ranches, drained the swampy alluvial soils and divided the land for landless peasants, creating a pocket of 211 family farms, of between 10 and 60 ha., in a region of big estates. In the 1970s the farmers produced cattle and rice, but the dry season made the rice-growing risky even with over 2,000 mm of rain a year, and in the 1980s a World Bank loan was used to develop irrigation. Farmers and agronomists maintain that the rainfall has declined as the forests have been cleared. In contrast to the remoteness, poverty, simple technology and difficulty of commercial production in San Lucas, El Distrito is part of the urban–industrial complex. Farmers eat very little of what they produce and for their high-technology production must rent or buy many essential inputs, from fertilisers, herbicides, pesticides, tractors, fuel, irrigation water and drugs for cattle to crop-dusting aircraft and combine harvesters. A third have tractors. By San Lucas standards, El Distrito farmers are rich. Unlike San Lucas, where men leave their own land every dry season in search of manual work on commercial farms elsewhere, in El Distrito, expensive technology is replacing manual labour. Many farmers live in basic comfort and several are prosperous enough to send their children to university. Nearly half the houses have electricity (some of it solar), many have electrical appliances and three-quarters have flush lavatories. Farm families are smaller and few labourers live on the scheme.

This high-energy, high-risk farming also produces a high turnover of farm families. We interviewed a random sample of 10 per cent of households and found that many people had been there less than five years. Rice farming depends on heavy borrowing from the bank, and many farmers, unable to pay off their loans, have had to sell. Yet each loan is evaluated in advance by an agronomist for the bank, and the agronomists say that failure is the result of bad luck, not bad farming. Land settlement in El Distrito is based on a technology so risky in this environment that farmers fail through no fault of their own.

La Payoa

Some 50 km south-west of El Distrito is a new co-operative inspired by Catholic and peasant leaders: La Comuna de Payoa (Townsend 1986). Here, violence on the ranches had reached the point where their owners inspected them only from light aircraft, without landing. In the highlands to the east, the (Catholic) Pastoral Service of Santander, SEPAS, and the tobacco and fibre-growing unions were struggling to promote consciousness, co-operation and grass-roots organisation. The NGO and unions saw land settlement as one solution to both the poverty of the highlands and the violence of the lowlands. The state agrarian bank bought 3,000 ha. of ranch land for some US$500,000 in order to sell it again (for the full purchase price) to a co-operative of fifty-eight landless families, organised by SEPAS. The move began in June 1986. We interviewed all forty-six families living in the co-operative in October 1987 (Townsend 1989, 1993a, 1993b). At that time, all the land had already been cleared for ranching, some of it rich alluvial soils, some badly eroded on the gravel hills.

The aim of the co-operative is to feed the members first and the local area second, with limited involvement in the national market. Access to the outside is much poorer than in El Distrito, but trucks can reach the community. Every commercial opportunity must be taken, however, because of the need to pay for the land: 'The serpent here is the debt we have to pay' (co-operative member). Because labour demands remain high, families are large, with many relatives brought from the highlands to work. Some 2,000 ha. are worked collectively (which is problematic) and the two tractors are centrally managed. Each household has 15 ha. to work individually. Incomes remain low, but living conditions are much better than in San Lucas because co-operative members have invested substantially in housing (though not in water, electricity or drains). Though there is no activity by guerrillas or death squads, life remains a great struggle as people try to contend with the unfamiliar heat and diseases and to learn new ways to live, work and organise.

Who does what and where?

These three small areas have very different economies and different work priorities. The differences are reflected in the demographics, particularly in sex ratios. Underlying these demographies are distinctive expressions of gender roles.

So many men

For us, there were two big surprises in the Serranía de San Lucas. One was women's roles as housewives rather than producers, as we shall explain later (p. 41); the other was the sex ratio. Among the adults (aged 15 and over) in the seventy-five households sampled, there were 189 men to every 100 women! No less than 60 per cent of the population is male. In El Distrito, the ratio was 153 men to every 100 women, and in La Payoa, 136 for every 100. More men than women had migrated in, and, as girls grew up unable to earn, they became too expensive and were sent to the towns, usually as domestic servants with some promise of education.

Nationally, in 1985 there were ninety-eight males for every 100 females in Colombia (Colombia 1986), but rural areas had more males (108: 100). Far more economic opportunities exist for girls in urban areas and they may also migrate to the towns to get away from rural life and women's particularly subordinate position there (Ordoñez 1986). The rural areas of Colombia with particularly high proportions of men are those of high immigration: rural Quindío has 171 males for every 100 females (Ordoñez 1986). Demographers refer to such patterns as 'the male pioneer model', which seems closely linked to a lack of economic opportunities for women. In the government colonisation schemes of the Dry Zone of Sri Lanka, for instance, there are less opportunities for women to earn than in the Wet Zone, so more males migrate in and a balanced sex-ratio is never attained (Kearney and Miller 1983). In India, David Sopher found that all areas of recent rural settlement had deficits of wives (1983). In Ethiopia, the lack of women in land settlements is reported as a serious problem (Chole and Mulat 1988). In the American West, the gender balance seems to have differed according to area and time: on the early mining frontier, women were greatly outnumbered in the early days, but in other areas where farm families were settling only about a quarter of the men were single (Jensen and Miller 1980). Even in Colombia today there is variety: in the Magdalena Medio, most of the 'excess' males are in the farm families, while in eastern Colombia, most are single labourers (Meertens 1988).

Housewifisation

This imbalance of sex ratios presents a puzzle. How have people built their gender roles and how are they related to this demography? As people have moved to the Magdalena Medio, there has been considerable change in what it is to be a woman or to be a man. On the whole, women have given up their productive roles and become housewives. It seems that they then become an expense which, in San Lucas, only about one man in two can bear, so that many girls and women leave. Therefore, caring for so many men and children becomes extremely arduous for the remaining women.

The 1980 Colombian census recorded that less than a third of those working in agriculture were women. This contrasts with the findings of Diana Deere and Magdalena León de Leal (1982), who had shown from fieldwork in Colombia and Peru that the northern Andes have 'family farming systems' in which all the family are involved in production, although women in more prosperous families much less so since agricultural work is undertaken by women only as a necessity. Our initial objective in this fieldwork was to demonstrate women's importance in production on the frontier. We therefore used a questionnaire similar to Carmen Diana Deere and Magdalena León de Leal, and were much surprised to find housewives, not women farmers (Townsend and Wilson de

Acosta 1987). Women proved to be almost excluded from the generation of income, especially in San Lucas (Table 1). Only one woman in these surveys, the head of a landless household in San Lucas, worked for pay in agriculture and not many even worked on the family farm. Women's work is in the house, bearing children and feeding and caring for children and men, full-time and overtime. As settlers, they have become housewives (Townsend 1993b). It is in domestic work that they can earn, in a village by washing clothes or selling cooked food, on the farms as domestic servants. Farm women's main contribution to family income (Table 1) is when they cook for the farm labourers (compare Deere and León de Leal 1982; Meertens 1993), so that the labourers can be paid less, at the expense of housewives rising at 3 a.m.

A woman's life here is defined in terms of her home, her children and sometimes her chastity; a man's in terms of his economic achievement and number of children. 'Sex' means heterosexual sex. Homosexuality is mentioned only when death squads threaten to kill 'communists and homosexuals'. In conversation, women talk of the needs of their family, of the farm, almost never of themselves or their own interests.

What, then, could we establish with questionnaires and informal interviews about women's lives in the Magdalena Medio? With hindsight, Janet Townsend and Sally Wilson de Acosta realise that they approached these studies in blinkers, as a socialist-feminist and a nutritionist. Our field notes deal mainly with the productive activities of the household, with illness and with nutrition. Questionnaires designed before the fieldwork (to be comparable with Deere and León de Leal 1982) ask the researchers' preconceived questions. Even in the informal interviews, we asked and women answered at the level of the

Table 1 Colombian surveys: selected tasks by locality; ratios between women and men who 'always do' the task (age 13 and over)

		Area	
Task	*San Lucas*	*El Distrito*	*La Payoa*
Cook for labourers	43:0	7:1	16:1
Laundry	10:1	8:1	18:1
Cooking	6:1	8:1	18:1
Hens	3:1	8:1	11:1
Fetch water	1:1	4:1	3:1
Pigs	1:1	7:1	15:1
Fetch wood	1:6	4:1	3:1
Garden	1:6	1:3	1:2
Milking	1:2	1:7	1:2
Agriculture	1:60	1:52	1:70
Cattle	0:65	1:28	1:13
[n:]	[337]	[116]	[141]

Source: Authors' fieldwork

household. At night, we would chat for hours and women enquired vigorously into our personal lives, but we felt it impertinent to ask about theirs. We did not learn the woman's-eye view. It was later that we found Robert Chambers' (1969) list of the problems of women settlers: title to land, access to income, workloads, isolation, all of which we can evaluate from our data.

Women in land settlement

Title to land

Title to land was given only to men in San Lucas and El Distrito, and, although wives and daughters have inherited and in La Payoa four households chose to have the farm in the woman's name, it is always a man who runs the farm. But women's position is contradictory. Women do not usually own land or contribute directly to production, yet women matter: several farms in San Lucas had been abandoned 'because the woman left'. Alfredo Molano (personal communication) thinks that women on the frontier are in an unusually strong position because the farm depends so much upon them. Donny Meertens (1993) reports that peasant women in eastern Colombia insist that their work has more recognition now from the men than before colonisation. We cannot agree, but Etilda, who has provided Justo (p. 21) with eight sons and sustained them all while they built a farm in San Lucas, is regarded by the family as co-founder and -builder with Justo. Staying has its rewards for a few. Women who have left, however, could take only their children, having no rights in the farm they had helped to build. As a woman at La Payoa said of her husband (emphasis added), 'I don't know how many cows *he* has.'

Access to income

In lowland ranching areas of Latin America, women often have little access to income as they are excluded from work with cattle or pasture (Hecht 1985; Court 1986; Lisansky 1979, 1990), although in the cool highlands cattle may well be primarily women's work (Deere and León de Leal 1982). In the Magdalena Medio, women and men say that the zebu breeds of cattle are savage and unsuited to women. Not all zebu cattle in Latin America are 'savage', for Donny Meertens (1988, 1993) found that in eastern Colombia women are involved in the clearing of the forest, the crop-growing (especially coca) and the raising and milking of the very same breeds of cattle. In the Magdalena Medio, cows (rarely bulls) do injure and kill and vicious cows make pastures into dangerous places, but their behaviour may relate to their handling rather than to their breed. Cows, like women, are socially constructed. But is it savage cows that keep women here out of the fields? Judith Lisansky (1979, 1990) found women in Brazil 'helping' on the pioneer farms although not on the ranches, but in San Lucas only three women out of 114 worked on the land at all.

Women in La Payoa (all new migrants) told us that moving there had cost them access to income. In their old homes, they had contributed to the clothing and feeding of their children from their earnings from craftwork, the selling of food and the raising and selling of small livestock. They have now lost all these opportunities and were in constant search of a new source of income. Carmen sells clothes and shoes on commission while others talk of learning basketry, embroidery or tailoring rather than work on the land. Most of their poultry and animals died on the move, but several still have pigs, which they sell locally to pay for household expenses, and chickens, raised mainly for home consumption. The men talk of the women rearing pigs in collective piggeries for sale to the city, but going out to such piggeries would be incompatible with women's long hours of work in the home.

Women in San Lucas also raise hens and pigs, with help from the men (Table 1, p. 42), but the only opportunities for sale are in the villages. A few women weave sleeping mats for sale, but in general women have little direct access to cash.

A small national scheme has reached El Distrito, which facilitates the generation of income by farm women by providing credit for the raising of their own animals. This assistance is offered for an activity which the man does not practise on the farm (two women had tried unsuccessfully to divert it into (his) farm activities). Eurania Rojas works with farm women for the Ministry of Agriculture. She promotes pig-rearing which can be profitable if food is grown for them on the farm. Esmeralda, for example, supports her children in school in the city, with the help of the pig money. Eurania also developed a successful programme of kitchen gardens with women, but crop-spraying has ended these as well as the men's small orchards.

Women's main direct contribution to household income remains cooking for the farm labourers (Table 1, p. 42). For most women, escape from poverty is a constant preoccupation. In La Payoa, many spoke to us of their need to earn, as they had in the highlands, but they sought skills to make money in the home, not on the land, and talked of livestock-raising, weaving, sewing and basketry. SEPAS (see p. 40) set up some training schemes for the La Payoa settlers, but these were strictly in the care of cattle for men, and in nutrition, health and contraception (by the rhythm method) for women. It appeared that SEPAS's ideas, like ours, were very much formed by men.

A striking feature of the co-operative of La Payoa is the blindness to the skills of educated women. Elena, who is better educated than any of the members, has come from the highlands to do the co-operative accounts, for a wage and as an act of social commitment. Unlike the members, but like the three young women employed for a wage by the co-operative as primary teachers, she works short hours and is extremely bored. The co-operative is desperately short of skills in accounting and in dealing with the bank and bureaucracy. Elena and the teachers are a considerable potential resource for education and organisation. The co-operative manager is a dedicated peasant leader with entrepreneurial skills and some knowledge of farming and cattle. He works long hours at the accounts and with the bureaucracy, but he and the co-operative council perceive young women as transient and unprofitable. Young women are not regarded as potential partners, let alone leaders, but as people who will do the minimum and leave. Elena has no place in La Payoa, nor have the more educated daughters of co-operative members, and all plan to leave because there are, indeed, no opportunities. One woman settler is an experienced practical nurse, but despite her large family, she is expected to offer her skills and time to other settlers free.

A woman and her children depend for their economic survival upon her sexuality, for their livelihood comes essentially from her finding a partner. Very few remain childless after their early 20s. In La Payoa, marriage is usual (in theory, it is a requirement for membership). In San Lucas, which is so much poorer, it is rare and many women change partners several times, sometimes with excursions into prostitution. Poor, uneducated women, with or without partners, may work as domestic servants in all areas, but their children tend to be the most malnourished and disadvantaged of all. Other women head for the towns.

Women's workload

As we did not ask, we do not know whether women's workload had increased with their move to the frontier. We know (Table 1, p. 42) that many girls are active in daily chores, such as laundry and fetching water, from the age of 7 (or earlier, in San Lucas). With very few exceptions, women are involved from childhood to old age in cooking for the household, cooking for labourers, cleaning the house, washing clothes and caring for children. After the age of 50 most of them do less of the heavy work, such as carrying water and wood, and more supervision. Women work much longer hours than men, for work in the house can be done by paraffin lamps or electric light. Many men spoke to us of the value of this work and in San Lucas, where women are particularly scarce, men join in to an unusual degree (Table 1).

Isolation

We do know that women are isolated. In all three areas, throughout her life a woman's archetypal role is in the house. She goes out mainly to do the laundry. Both forest and pastures are wild and 'dangerous' for women and children and even in the garden it is mostly men who do the work (Table 1). Girls learn their work in the house, boys in the fields. In all three areas, women say that in their home villages they felt safe to move freely, but here they live among strangers. In San Lucas and even El Distrito, many do not know their women neighbours,

45

and in San Lucas some had had to deliver their own babies. Men settlers leave the farm frequently, to buy and sell (in San Lucas, the men do the shopping), to find paid work and to drink with neighbours. If men do not build a social life beyond the household, the farm will fail. Women have no such outlets, and come to value seeing only the family 'because people are so horrible' (a woman at El Distrito). The settlers at La Payoa are the newest and the women the most vocal, having just lost their social world of mother, female relatives and perhaps school friends. La Payoa women are also the least alone, for when they arrived most of them lived in groups in the big old ranch houses while their own houses were built. Though this time was a crowded and miserable experience, it did at least mean that they got to know each other. The women complain now of the long walks and dangerous cows, but many take their children to the Sunday co-operative meetings which are major social events and many also get to women's meetings or even visit each other. Elsewhere, the escape for prosperous women lies in their children's education. Three of the seventy-five San Lucas households have a house in a village where mothers and children live during the school year so that the children are able to finish their primary education. These houses have electricity, running water and at least a latrine. El Distrito families must in theory live on their farms under the land reform rules, but it is recognised that their children can only go to secondary school if the family lives in town, and several do. One woman runs a business in the town selling milk from the farm.

Women's lives

Domestic labour

Wealth and poverty make little difference to the work women do in the house but a great deal to the conditions under which they do it. In San Lucas, houses are poor and lack water, electricity and sanitation. Fruit grows in the garden but few vegetables, not even beans. Milk and meat are seen as the ideal source of protein although only the prosperous have milk and, for most, chicken is a rare luxury. If there is a school, it offers only the first year or two of primary education, so that illiteracy is widespread. Doctors are scarce. Papayal, the village in which we worked in San Lucas, has a health centre that employs doctors, a dentist, dentist's assistant, midwife, paramedic and hygiene adviser, but, more often than not, they are absent in the coastal cities, two days' travel away. (While we were there, a young and penniless woman went into false labour and we were asked to take charge, as the only medical person actually in the village was the dentist.) Those who can afford it travel for days to doctors in the cities. All women toil long hours to care for their families in these conditions, although more prosperous women can hire resident help.

Living conditions are much better in El Distrito, the land reform case-study, but health is threatened by new environmental hazards from the high technology. Most houses are of brick or cement block, with tin or asbestos roofs, cement floors, running water, lavatories and even gas to cook by. Schools have all primary grades and most adults are literate. Health services are available. We found most children have been vaccinated and buildings sprayed regularly against malarial mosquitoes. But the use of pesticides has been introduced without adequate warnings or precautions. It is seen as tough and masculine, for instance, not to bathe after using these chemicals. The pesticides, being valuable, are often stored under the bed. Many homes stand in the middle of rice fields and, according to Eurania Rojas of the Ministry of Agriculture and Dr Quiñones (director of the nearest hospital), people are constantly poisoned, because their water tanks are open to the sky and so to aerial crop-spraying. Farmers complain that fruit trees and chickens die, and, when neighbours forget to warn before spraying, so do cattle.

At La Payoa, SEPAS and the co-operative members set a high priority on housing, health and education. Most adults are literate, most children are vaccinated and malaria spraying is up-to-date. Members have incurred substantial debt for housing, even using some money lent for crop production (which has led to repayment problems). Even so, many basic services are lacking, including clean running water, electricity, latrines and medicines. In an emergency, for example, a man injured by a cow or a woman in problematic labour, the hospital is several hours' drive away on a tractor.

Migration

Every life in the Magdalena Medio is coloured by migration. The pioneers arrived as migrants but some children leave home for education, many young girls and women leave for the cities and many men must leave home frequently to find paid work. Some of the settlers had travelled widely, to zones of commercial agriculture across Colombia, to the cities and perhaps to Venezuela. Many men but few women return, although, especially in San Lucas, women may bring their children back to grow up on the farm. As a result, it is normal for adults to have brothers, sisters and perhaps children scattered across the country, engaged in activities ranging from domestic service and casual labour to studying at a university. These networks provide an important resource, especially for the better off, for placing children and grandchildren and for support in case of sickness or old age.

CONCLUSION

In 1984, we set out to explore what we thought was the unacknowledged importance of the productive farm work of pioneer women in San Lucas. We found their importance to be as housewives, in contrast to our expectations from studies in highland Colombia (Deere and León de Leal 1982). Not only do they become housewives, they do so at the time they arrive, not when the cattle take over as in the Mato Grosso (Lisansky 1979). Very few women work in the fields, unlike eastern Bolivia (Hamilton 1986), and very few with cattle, in contrast to eastern Colombia (Meertens 1993).

We learned something of the hardships of pioneer life and of the women's struggles to protect their families from malnutrition and disease in the face of the overwhelming need to establish the farm so that few resources are left for survival. We found children who weighed only half what they should. By publishing our material, we were able to make women and their suffering visible.

In 1987, we found differences in the middle Magdalena valley between places with contrasting types of state intervention. San Lucas has low-cost 'spontaneous' colonisation under which state land was given to pioneers, but which suffers from a lack of state support in services, technical advice and credit. The hardships, poverty and isolation bear heavily on women. El Distrito has high-cost, high-tech land reform which is not only too costly to be extended but very 'top-down', promoting risky technologies which cause many farmers to fail. Women whose farms do not fail enjoy good conditions but are isolated. NGOs are now extremely popular as development agents (Bebbington 1993), and La Payoa illustrates some of the problems and advantages of NGOs. There, SEPAS has set up a co-operative which combines good intentions with grinding debt, but at least the women know each other and can call on a network of resources.

Our survey data revealed the predominance of males which we had, oddly, not recognised in the field nor heard mentioned. It also enabled us to make detailed comparisons of the gendering of tasks, age structures and living conditions between the research areas (Townsend 1993a, 1993b) and to supply reports to government bodies based on survey evidence confirmed by observation and oral information (Townsend 1989). Such background information would also be essential as a context to, say, life histories.

There are still many themes which we cannot explore through surveys, even with informal interviews. We have little understanding of gender relations. We heard at length but at second hand of drunkenness and domestic violence, both always 'other people's problems'. Similarly, we are not confident that we know what women's priorities are. When we asked them, they spoke on behalf of their households of the need for income and services but could rarely specify what they wanted for themselves or their daughters. The questions were too unfamiliar and were asked of each woman by an outsider, not discussed among a group of local women as we were to arrange in Mexico (Chapters Four and Five). Similarly, we tried unsuccessfully to learn of relations between men and women.

It is clear that women have little formal power, but we do not know how significant this is, and many older women have considerable informal power. In La Payoa, Carmen is one of the most powerful people in the community. Her husband was a union activist in fibre production in the highlands and is now a co-operative leader. Of their eleven children, five live with them, one is in the city and five others are full independent members of the co-operative, as arranged by Carmen. At co-operative meetings, Carmen's is one of the most respected voices. Etilda (see p. 43) and Carmen have succeeded by using existing institutions, not by challenging them.

We must now consider the concept of practical and strategic gender needs, currently so controversial. Maxine Molyneux (1985) showed that women in a given place often share specific practical and strategic gender interests. Women's practical interests lie in the fulfilment of their existing roles, while strategic interests focus on changing their position in society. Caroline Moser (1989, 1993) has developed the concept of gender needs. Pioneer women, for instance, express a practical need for clean water, but not because only women need water. Rather, women feel and express the need because it is they who must provide water for others. It is they who are the caretakers of health for their families, and they who, in El Distrito and La Payoa (Table 1, p. 42), carry most of the water. Some women, however, might wish also to identify strategic gender needs in order to change society and their roles within it. Such strategic gender needs might be paid employment and land titles for women or rights against violent husbands.

It is necessary to mention here the many problems with this concept of difference between the strategic and the practical (Alsop 1993). The distinction is like that made between the formal and informal sectors or, by feminists, between public and private space or production and reproduction. All are like salt in being valuable when appropriate. (A low salt consumption is healthy in cooler climates, but will produce severe muscle cramps in the heat of the middle Magdalena valley.) Distinctions between practical and strategic, formal and informal, public and private, production and reproduction all relate to our habitual ways of thinking in the West. These divisions, therefore, have their use in thinking and especially in teaching, although all are problematic in theory and practice and can be thoroughly confusing in cross-cultural research. As one way of beginning to think, they are useful, but as the only way of thinking they are disastrous.

The concept can help us think about women's needs in the middle Magdalena valley. In a single interview with a stranger, women asked about their needs reply with the needs of their family (income, schools, clinics) and with some practical gender needs such as clean water, but not with needs which we expect such as rights in the farm or safety from marital rape. It is not, however, for us to pronounce on their wants and needs, since outsiders could readily, for instance, erode the bases of women's informal power while making no real improvements. It is often very easy for outsiders to get women beaten up. Positive change is much more difficult to achieve when information is limited to questionnaires and interviews, however pleasant. Simply, in practical terms, etic, outsiders' accounts built up on short visits without a determined search for emic, insiders' views are of extremely limited use for planning purposes.

4

WOMEN PIONEERS IN MEXICO: OUR ANALYSIS

INTRODUCTION

In 1990 and 1991, we sought to learn about women's lives as pioneers in Mexico. In this chapter, we shall recount our analysis of their situation, drawing particularly on what we learned from questionnaires and interviews in a random sample of households. This chapter therefore depends more on 'harder', more extensive information (Sayer 1984) which is either widely agreed or with which other researchers using the same survey methods would concur. In Chapter Five, we shall use 'insiders' voices', the words of pioneer women taken mainly from life histories, to describe individual lives. This distinction can only be very loose, but there is a real difference in the kind of information obtained. First, we shall briefly describe the lowlands of south-east Mexico, together with our research projects and the methodologies used. We shall then compare pioneering for women in Colombia and Mexico and show how pioneering in Mexico has led to local agrarian crisis. Finally, we shall briefly describe the diversity and differences in the fourteen communities in which we worked and describe the solutions that we, as outsiders, would wish to see to the problems we identify.

THE LOWLANDS OF SOUTH-EAST MEXICO

Much of Mexico is dry or even desert and most lies high above the sea, but the lowlands in the south-east are naturally clothed in tropical forest, even rain-forest (Map 1, p. viii). As in Colombia, the wet tropical lands had been densely peopled from about 3000 BC (Nations and Nigh 1980). For all the remarkable biological wealth, the arrival of the Spaniards brought European diseases, conquest and economic disruption which almost emptied these wet, lowland areas of people. In colonial and then independent Mexico, the population eventually recovered and grew, but the growth was mainly in the mountain basins of the centre and south-east. The highlands of the south-east were densely peopled, but in the forested lowlands only the Maya of Yucatán achieved dense rural settlement. Elsewhere in the lowlands, pockets of cacao, ranching or logging supported settlements. According to Fernando Tudela *et al.*

(1989), rich and poor in the lowland state of Tabasco seem to have had easy access to an adequate diet until twentieth-century banana plantations, oil production and land settlement brought 'deteriorating development', malnutrition and mass poverty. We believe this 'deteriorating development' to be widely characteristic of the south-eastern lowlands.

Railways and, later, roads brought exploitation to the forests, beginning with the railway across the Isthmus of Tehuantepec in 1907. The Mexican Revolution of 1917 gave all Mexicans the right to own land, but land reform was slow and its impact on the tropical forests delayed. Rather, massive state investment was concentrated in the dry north, in irrigation and irrigated land settlement, resulting in the 'Mexican Miracle' when Mexican agricultural production grew by 5 per cent per year from 1925 to 1965. In the south-east most new settlement before the Second World War was spontaneous and private, with low productivity and technology, as at Jasso (see p. 67). After 1940, Presidents Camacho and Alemán promoted the 'Advance to the Sea' to occupy the tropical lowlands. From 1946, in theory, the landless could move to 'unoccupied' or 'national' land in the lowlands (in practice, sometimes occupied by indigenous people or others). There, the landless could form *ejidos* (land reform communities) and claim plots of up to 20 ha. for each family. This occurred at Balzapote, La Corregidora, El Arroyo, La Planada and El Tulipán, where we were later to work. Private colonists could cultivate up to 300 ha. of crops (as at Sor Juana Inés) or create ranches of up to 2,500 ha. (as we found in Chiapas and Tabasco). In practice, the new laws gave landowners access to the thinly peopled lands of the tropics without producing the intended reduction of the pressure for land reform in the densely-populated areas (Revel-Mouroz 1980). There were official land settlement schemes, but it was not until the 1960s that land settlement became an important part of government policy, and even then most colonisation was spontaneous (see Cuauhtemoc, Plan de Ayala, pp. 65–7). In the 1970s in particular, the dream was to repeat the 'Mexican Miracle' by using big investments and modern technologies to occupy the tropical lowlands, as in Plan Chontalpa or Plan Balancán–Tenosique (p. 67). The dream failed. Neither the technologies nor the administration were appropriate, so all the schemes destroyed the forest but none fulfilled their objectives. Nevertheless, much forest was cleared for ranches and many landless people did achieve new *ejidos*. Many of these communities provided cheap labour for private farms, since people can grow some of their own food but must also find paid work. By the 1980s, there was little forest left and most new colonisation was in the Selva Lacandona (especially in Marqués de Comillas), in Campeche (including Independencia, Naranjales and Tacaná, see pp. 70–3) and in Quintana Roo.

This pioneering has been destructive in several ways. First, almost 90 per cent of the tropical forests of Mexico have been cleared (Leff 1990). Second, bitter conflict has arisen as ranching destroyed the forest and restricted the access of poor people to the land across most of the wet tropical lowlands. Third, in

51

government-organised colonisation, many people were moved against their will. Nowhere was there effective consultation, although efforts are now being made. Fourth, indigenous groups have been displaced and exploited (Revel-Mouroz 1980) although not directly by any of the communities which are featured in this book. Now, government policies have changed so that extensive restrictions on felling the forest mean hunger for the poor and lost profits for many of the prosperous (see p. 67).

Most agriculture, cattle production and forestry in south-east Mexico are unsustainable. High-technology initiatives such as rice production at Uxpanapa have failed (Ewell and Poleman 1980), and ranching is extensive and environmentally destructive. So far, new initiatives in tree-planting and aquaculture have had little success. The lack of economic, sustainable systems is now familiar across much of the humid tropics, in contrast here with the period before the Spaniards came.

In December 1991, after our fieldwork in these communities had been completed, the 1917 Mexican constitution was altered so that not all Mexicans now have a right to land. The distribution of land by the government is essentially over. Until 1991, 'reformed' land was held by the community, the *ejido*, and could not be sold, but only granted to individuals, *ejidatarios*, to use and to leave to their children (although, in practice, it was often sold to incomers with community approval). Now, given the consent of the community, farmers may enter into joint ventures with private capital or may even sell their land. Therefore, the shape of rural Mexico will inevitably change.

OUR RESEARCH PROJECTS

We started work in July 1990, when Janet Townsend and Jennie Bain carried out a pilot project in Los Tuxtlas (Veracruz) and Uxpanapa (Veracruz and Oaxaca) (Townsend with Bain 1993), using public transport. In 1991, Ursula Arrevillaga, Socorro Cancino, Silvana Pacheco, Elia Pérez and Janet Townsend worked on the main project, in Los Tuxtlas again and near Palenque (Chiapas), Balancán (Tabasco) and Escárcega (Campeche) (Townsend *et al.* 1994; Map 1, p. viii). Fieldwork in 1991 lasted from June to August, using purchased vehicles to save time. It is important to note that the enterprise became Anglo-Mexican and was run on a collective basis (see acknowledgements). Janet Townsend's plan had been to work together in the first two communities and then split into pairs, but the Mexican authors were doubtful of the safety of this for five women, and so we all worked for about two weeks in each community rather than in pairs in the same communities for four weeks. (The exception was La Villa, where Janet Townsend missed the work but was able to meet some of the people.) As it was normally not possible to interview after dark, most evenings were devoted to discussion among the authors and to the checking and editing of the work.

How representative?

Our aim was to learn from pioneer women about their problems and the solutions which they proposed. Two misfortunes limited what we were able to do. First, two of us caught typhoid, which meant special diets for weeks, so we could not sleep and eat in the communities as much as we wanted. Then came the conflict over timber in Marqués de Comillas in July 1991. The government had banned the extraction of any timber from the area, even trees felled before the ban. When it sought to collect the timber already felled, apparently to sell on its own behalf, the pioneers rebelled and took hostages. Some even set off to see the President of the Republic. (We saw them in Palenque, men and women on trucks, singing and waving flags.) They were ambushed by the military further up the road, imprisoned, beaten and raped. We felt that this moment was not propitious to seek confidential information in Marqués de Comillas, so we lost our chance to work in the area of rainforest with the most recent colonisation in Mexico.

The women's voices in this book cannot represent all women settlers in southeast Mexico. The region has four forms of land tenure: the *ejido*, or land reform community, the *colonia*, or colony where land is privately owned, the *comunidad*, where land is held by an indigenous community, and the *pequeña propiedad*, the private 'small property'. None of us speaks an indigenous language, so, as we felt it important to concentrate our limited resources on hearing what women had to say to us directly, we worked only in Spanish-speaking *ejidos* and *colonias*. We did not go to *pequeñas propiedades*, because their workers fear that talking to outsiders could cost them their jobs. We suspect that such women may be as painfully isolated as those in Colombia. In new indigenous communities, too, problems and opportunities may be very different, so we look forward to the publication of current work by Lourdes Arizpe, Magalí Daltabuit and Xochitl Leyva with both indigenous and Spanish-speaking women in south-east Mexico. (Writing in April 1994, it seems that new indigenous communities which have lost their land to ranchers may possibly be an important source of recruits to the current Zapatista rebellion in Chiapas.) However, indigenous women pioneers do tell their collective story in the book–cassette *Sk'op Antzetik* (Calvo *et al.* 1992) in Tzotzil, Tzeltal and Spanish. They moved into the mountain forests at their husbands' command, against their own wishes and under great hardships. They record their early sadness in the heat and the thick forest, but now they 'work hard but eat well'. Their tale is one of work and pride: the pioneer dream. The tales we heard were also of work and pride, but of fear and desperation, too, as livelihoods disappear and women seek ever new sources of income to feed their families.

We feel that the women who talked to us represent women in hundreds of Spanish-speaking communities formed in the last fifty years in the tropical lowlands of Mexico, in Veracruz, Oaxaca, Chiapas, Tabasco, Campeche and perhaps Yucatán and Quintana Roo. Many stories and needs are very familiar from other parts of rural Mexico.

Methods

The greatest single restriction on our work was our limited time: only fifteen woman-months, divided between six weeks in 1990 and three months in 1991. Our selection of communities from the five relevant states was pragmatic. In 1990, we wanted to study the forest gardens of Los Tuxtlas, Veracruz (see p. 62), on the more accessible Caribbean coast, and the well-documented land settlement of Uxpanapa (Veracruz), in the remote centre of the Isthmus. When we could not arrange to stay in the actual scheme of Uxpanapa we worked nearby instead (in Veracruz and Oaxaca). We used questionnaires and unstructured interviews and collected our first two life histories, one of which was taped. In 1991, we were a new team and therefore went again to Los Tuxtlas (to different communities) in order to begin again with the forest gardens and to develop our methods collectively under easy conditions. We did not go again to the Isthmus because of news of frequent attacks there on vehicles moving between communities. In other states, we selected communities around Palenque (Chiapas), Balancán (Tabasco) and Escárcega (Campeche) by using field visits and information from land reform officials. We had wanted to study an *ejido* and a *colonia* in each state, but could not get the necessary information to do so.

In Mexico, unlike Colombia, people are more comfortable if researchers have official recommendations, and permission from the elected village leader is important. We took letters from the Colegio de Michoacán (our academic affiliation in Mexico) to municipal authorities, who recommended us to community leaders, all of whom were extremely helpful. We sought *key informants* among land reform and agricultural extension officers responsible for the communities, and among community leaders, health workers and midwives. Since much of our work was in the summer holidays, teachers were frequently absent.

Over the two years we conducted *questionnaire surveys* in twelve communities in five states, in households selected at random, to establish living conditions, economy, demography and divisions of labour by gender and age. (In these areas, village leaders hold accurate lists of households so that random sampling is easy.) The questionnaire used in Colombia was modified by the Mexican authors for crops, land tenure and building materials. After administering each questionnaire, we conducted an informal interview if possible, and acknowledged the help with an 'instant' photograph of the family (worth about a day's agricultural wage).

Only one household out of 241 sampled refused an interview – on a Sunday, in pouring rain. There were important differences from Colombia in terms of the responses received. In Colombia, people were happy to say, for instance, roughly how many cows they had or what area of what crop. In Mexico, these questions seemed threatening and were often answered untruthfully or not at all. In Colombia, people gossip about their neighbours' farms, while in Mexico this seems improper. Conversely, Mexican women pioneers speak more freely about their personal lives, possibly because they are less isolated.

In each community we then asked some women to tell us *the story of their lives*, to give us an insider's view of what it is to be a woman in these communities. In identifying these narrators, we sought a variety of attitudes and experiences combined with a willingness and a capacity to talk. Only one woman out of the twenty-seven asked was 'too busy'. All were happy with the cassette recorders once the family had had fun with them. Above all, we wanted to find out what the narrators themselves wished to say. We believe that, because of this relaxed approach, we know what they wanted to talk about, but we also know that we lost a few migration dates in the process (because we failed to ask for them) and that the transcribed narratives seem shapeless.

At the beginning of our work in 1991, we developed a checklist of points to be discussed if possible, and endeavoured to use this list as much as we could. In each area, we recorded life histories, transcribed them in manuscript, checked them and, when we reached a town, copied them and posted them to the Colegio de Michoacán for typing. We would have done better to have taken word-processors into the field, since typed life stories are so much easier to check. We did fail to recognise some serious problems, such as one failure to ask the place of birth, unnecessary leading questions (see p. 140) and issues of balance. A further problem was that our interests were not uniform. For instance, Janet Townsend's prime concern, having studied colonisation for many years, was with the process of settlement, the reasons for the move and the experience of pioneering, while the Mexican researchers, not being geographers, were more interested in social issues. Elapsed time was a problem as most of the communities were more than fifteen years old and so women had to try and recall moves made long ago. They gave us repeated accounts of flora, fauna and hard work in the early years but we should have pressed more for personal memories. It was the Mexican authors who suggested asking about first menstruation, and we shall see (pp. 88–9 and Chapters Eight to Eleven) that many women not only had a great deal to say about it but then talked more freely about their personal lives. We also initiated questions about sexuality, and Silvana Pacheco was particularly successful at asking these in an acceptable way. She was involved in recording three of the life stories published here (Chapters Nine, Ten and Eleven). Most life histories were recorded by two of us and transcribed by one, although some were solely the work of one person. We all participated in this task which we found very satisfying if sometimes painful. Our attitudes differed as individuals, with some of us seeing social scientists as almost entitled to intrude while others were content with what narrators wished to say, but the differences were in detail. Not all of us would ask the very personal questions but none of us would push them when they were unwanted.

Where possible, we also arranged women-only *workshops*, at which the women told us about their problems and proposed solutions. (We shall use 'workshop' as the literal translation of *taller* and because 'focus group' is not yet widely used in development studies.) In Tacaná, we invited all the women in the village, through the (male) village officials, and some 150 came; elsewhere, the Agricultural and Industrial Women's Group (UAIM) called the meeting for us. Under Mexican agrarian reform law, each *ejido* is asked to set aside one plot of land, of the same size as the family plots, for 'landless women'. Such women must constitute themselves as a formal group to use the land, and are then, if they work it collectively, eligible for credit and training. Non-agricultural UAIMs are rare, and we met none. Members of these groups include the most needy and some of the most powerful in the village, for the qualification is landlessness, which includes the wives of *ejido* members and their children as well as the near-destitute. The publicly active and the wives of community leaders are often leading members. The range of income and opinion is wide, as is that of age as often both mother and daughter join. Other members of the community did attend our meetings. Thus, while not representative samples, the women who came to workshops did bring a great range of opinion.

We conducted six workshops (Tables 3–5, pp. 66–9), including two in La Corregidora by request. At each workshop, we would begin with an 'ice-breaker' when two of us would be tied together and would attempt to eat biscuits from well-separated plates, heehawing loudly in frustration. Eventually, we would co-operate and converge successfully on the same plate, to laughter and applause. One of the Mexican authors would then explain our presence and arrange for everyone present to introduce their neighbour to us all. She would then form groups, separating those who were already together and ensuring that each group had a literate secretary, before asking each group to record their reactions to the question, 'Do I like to be a woman? Why? (Or why not?)' Each group would report and we would summarise on wall charts. We would then ask each group to consider: 'What problems do we have as women in (name of community)?', and then, 'How shall we solve these problems?' Again, each group would report and we would put the answers on charts. General discussion would follow, with complimentary fizzy drinks from us. The procedures were unfamiliar to most and we would all usually have to move around and help. We tried not to introduce questions, but may sometimes have failed in this endeavour. We directed the proceedings, but tried not to influence what was said. Afterwards, we wrote out the questions, answers, discussion and conclusion for each workshop from the charts and gave copies of this to the group. These occasions normally attracted twenty to thirty women for two hours or more, and were very popular.

Before leaving each area, we wrote up general accounts of its communities using our interview notes, workshops and life stories. These were given to the community leaders and relevant local officials once they had been typed.

Unexpected developments

In 1990, pioneer women changed our minds, through our sixty interviews and two life histories (Townsend and Bain de Corcuera 1993). Janet Townsend and

Jennie Bain had come with a very 'socialist-feminist' agenda, much concerned with women's production and access to income through, for instance, home gardens. Pioneer women proved more concerned with alcoholism, domestic violence and marital rape, although expenditure on alcohol is also a highly economic issue. Both life histories dealt at length with alcoholism and one with violence and rape, which changed our focus on pioneer life. Alcoholism also came up during the unstructured interviews after the questionnaires. It may be important that although Janet is British and Jennie is Mexican, we are both fair and look foreign. In three of the four communities we visited, we were asked for some drug to put a man off alcohol – not one for the man to take himself, but something to be slipped into his coffee or his food, like a love potion. We were educated, foreign-looking women representing modern, technological society who had access to the high technology which produced contraceptives and so surely could produce a drug against alcohol? To Janet's surprise, she was not asked this question in 1991, nor were the other Mexican authors. Possibly our arrival was less strikingly foreign with a majority of Mexican-looking interviewers. Even so, in 1991 alcohol was a frequent topic of discussion. In 1990 the women had shown us the sexual conflicts around reproduction, provoked at least in part by agrarian change. Thus, in 1991 we introduced the question in unstructured interviews and, if necessary, life histories. The issue also came up, we think spontaneously, in all the workshops in 1991.

A team of researchers?

We tried to work as collectively as possible, seeking to confront the difficulties, but of course there were differences among us. For instance, we come from very different cultures of work. England in the 1990s has both much higher rates of pay and a higher rate of exploitation of professionals in the public sector (through very long hours of work) than Mexico. These hours could hardly be asked of Mexican professionals, paid at Mexican rates, and British rates would be unfair competition to Mexican researchers wanting field staff. A tendency to defer to Janet Townsend even when others knew better had to be overcome. Our skills, interests and experience initially varied greatly. Only Janet Townsend had worked in pioneer areas, only Silvana Pacheco had conducted workshops (*talleres*) in Spanish, only Ursula Arrevillaga had extensive experience of life histories – which she had been taught to transcribe in the third person, editing out repetition. (We agreed to transcribe verbatim.)

Why was this project possible?

Why were women and men so willing to talk to outsiders, one of them a foreigner? Essentially, their responses reflect the context of Mexican rural communities which depend for all their services on negotiation with outsiders, bureaucrats and politicians, from the mayor to the President of the

Republic. This is how they get land, teachers, a school, perhaps a road, electricity, water, drains, a clinic (or at least a health worker) and any outside advice or credit. They must work on the interface (Long and Long 1992) between the community and the outside world, and must do this mainly by going out, to town or city, as even agricultural extension officers rarely come to the community. Dorien Brunt (1992) has shown how skilful a community can be at this manipulation, and how this operation in the public sphere tends to be limited to men. Much of the informal manipulation, such as going to the town and taking bureaucrats out to dinner and perhaps to brothels, is open only to men. Women tend to be restricted to the official, formal channels, and therefore to be far less successful in managing this interface.

We think that we were perceived as an opportunity. As educated women, we represented the power elite, outsiders suddenly accessible. In most of these communities, no woman with a degree had ever been seen. Even the women officials whose jobs are to support the women's production groups (the UAIMs) hardly ever go to the communities. Although we explained that we had no resources, no projects, no credit to offer, and that we planned only to publish a book about women's lives in these communities, we think that, as outsiders, we were seen as a resource and a way of getting their story told. There was suspicion, of course, and we were frequently interrogated. In Jasso, there was a rumour that we were Americans 'coming to take the land' and, in Sor Juana, the men were searching for us in the dark with shotguns after a rumour that we were robbers with machine guns. In each case, the community leader allayed suspicion. Many women set out to manipulate us and probably succeeded in creating the impression they intended; this was the price of acceptance.

We were surprised that the men were so supportive, all the more so when we heard how critical the women were of them. The men were friendly and helpful. In Tacaná, for example, which is 9 km from a road fit even for a jeep in the dry season, the village leader and his committee, all men, arranged to meet us at the road with horses to carry our hammocks and food, and lent us the teachers' hut to sleep in. All the communities are in economic crisis and, in all of them, many men want women to earn more and, of course, want better services. We think that we and our written studies were seen as ways of bringing attention to their problems and plight.

MEXICO AND COLOMBIA: PIONEERING FOR WOMEN

Rural Mexico is a big surprise after Colombia. The politics are very different from the rest of rural Latin America, for the same political party has been in power for sixty years, and in all the villages we visited the government has extremely well-organised political support and, notionally, duties. Tight structures of local government reach every village through a local municipal agent, almost always a man. Because government and opposition, national and local, are overwhelmingly male, women's groups often turn to the wife of the state governor when they need help.

The lives of Mexican women pioneers are very different from those of Colombians, for Mexican pioneers live in villages, not on isolated farms. If a group want title to public land as an *ejido*, they must live in a village, and even in private *colonias*, most farmers live in the village with their plot perhaps an hour or more away even by truck. Only landowners and indigenous people are likely to live on the farm.

In Mexico, as in Colombia, a woman's relationship with her mother is strong and enduring and its loss following the move to a new settlement or to a city is a great pain, but Mexican pioneers at least have neighbours. Pioneer households in Mexico are smaller than those in Colombia. As Mexicans live in the village, it is much easier to hire labour and less necessary to house it. In Mexico, more than two-thirds of the families in the surveys consisted only of parents with their children while in Colombia, less than half were nuclear.

The demography is different too. In Colombia far more pioneers are men than women, but in our surveys in Mexico there were nearly as many females as males at all ages, although among those aged 15 and over, there were 112 men per 100 women.

Divisions of labour by sex are less extreme in Mexico, but quite similar by age. In Mexico, a few more women join in agriculture and ranching, but the biggest difference is that everyone, but particularly women, participates much more in work in the garden. However, living conditions are very similar in both countries.

FROM PIONEERING TO AGRARIAN CRISIS IN MEXICO

All the communities in which we worked are new creations and physical access to markets is still difficult. The pioneers created communities out of the forest through many years of great hardship. Now their survivors among the old people can say, 'My children don't suffer as we suffered' (Cristina, Cuauhtemoc). But all are now facing the same agrarian crisis which transforms women's lives: the change from food crops to cattle. In the wet tropical areas of Mexico, 'the cattle are eating up the people'. As soon as there is no more forest to clear, the settlers sow pasture.

The communities in which we worked had all been spontaneous settlements. Often, men had come to inspect the new area, and had then brought their families. The pioneers would live in the forest, perhaps in a shack of palm leaves under a tree, 'sleeping in hammocks or on the round poles of saplings'. Privations were severe and often they had to divide one tortilla (pancake of maize flour) between two adults for a meal. Snakes and biting insects were daily problems. Markets or medical attention were at least a day's walk away so infections bore heavily on children, and many died. In some communities, 'the men felled, the women sowed' while babies slept in hammocks slung from the trees nearby. In others, women were fully occupied in the house. The pioneers began by living from their crops, sowing maize, beans, gourds and chillies in the ashes of the forest. A few products were carried for hours on their backs to market to buy salt, machetes and cloth. Eventually the settlers would seek recognition as an *ejido* or *colonia*. The role of the women was to sustain the men who worked in the forest and fields, to provide sexual services and, above all, to create the future labour force, for labour was a great asset as long as there was forest in which to make *milpas* (clearings to grow maize).

Now, the forest is running out, and crops need expensive fertilisers. Land use is changing from cultivation to cattle-raising, which offers very few jobs. This change has been described as 'ranch-isation' (Toledo 1990), the 'grass revolution of the American tropics' (Parsons 1976), 'hamburger and frankfurter imperialism' (Feder 1982) and 'deteriorating development' (Tudela *et al.* 1989). Many people are forced to go to the cities or to new areas in the forest. We met Alvaro when we interviewed his family by the Gulf of Mexico, and ten days later we met him again, in the middle of the Isthmus of Tehuantepec where his brother was once more 'opening up the forest'.

This agrarian crisis of the grass revolution means that there is far less work to do. Even in the *ejidos*, this is a time of division into rich and poor, essentially those who have cattle and those who do not: three-quarters of the families we interviewed did not. With the crisis, investment and education become important and traditional workers have few opportunities. Intensive cattle production, for instance, requires improved stock, more fencing, better pastures and much more rotation of pastures. It means calculation, investment and much negotiation with experts and with banks or other sources of credit. Education is then a great advantage, but on the whole people in these communities have had little time or even opportunity to obtain an education. Few people can afford to send children to live and study in the town and it has been more important to send boys to the fields than to secondary school, while girls are 'needed at home' and the long walk to the secondary school has deterred many parents from sending their daughters. Many of the older generation and even some of the younger cannot read and very few have enough education to develop a profitable modern farm.

The agrarian change also brings a family crisis. Far less labour is needed, paid work becomes scarce and many adults become superfluous. Only half of our survey households lived mainly from the farm, while half lived mainly from wages. When there is no forest to make *milpas*, money must be earned, food must be bought. Women have to confront radical change. In the pioneer communities many women worked on the land, but their crucial role was to maintain and reproduce the labour force. Skills were learned mainly in the family and traditional families were large, but the need now is for a very different labour force as household prosperity depends on new skills and far fewer hands can be used. Women were skilled in meeting the old needs but not the new, most having left school 'because my mother was ill'. Painful demographic change is in progress. Many young couples plan to have only two children (usually the man's decision) and many older women are sterilised. At the same time, many women in their 30s already have eight or ten children and want no more; by contrast, their husbands see contraceptive measures as a threat to their masculinity (see Lucia, pp. 107–8).

Women's role in general has been, they say, to 'help': to help men work by caring for them, to help children grow up to work and to 'help' make an income. Much of this activity is seen as pastime rather than work, for the real work is the men's (Table 2, p. 63). Most men but few women work regularly on the land. Men have much more leisure and very few contribute regularly to work in the home. Women's work is much more diverse, including cooking, child care, laundry and housework on the one hand, and earning money on the other.

In this crisis, everyone wants extra income. Even women want to earn, but the more rural the settlement, the fewer the opportunities for a woman. The smaller and the poorer the community, the less likely it is that any local woman earns a living, even as a midwife, and the more difficult it is for her to contribute to family income. She will produce eggs, poultry, perhaps pigs for sale, but beyond that most opportunities are in paid domestic chores: making tortillas in other people's houses, cooking for the elderly, doing other people's laundry, cleaning for others, making snacks for sale, selling door-to-door, working in shops, sewing for others or delivering babies (see p. 101). In a poor village, the whole of the paid work available adds up to very little, certainly not enough for a girl with secondary education, and there may not even be a full-time shop. In a town or more prosperous village, there is more demand for these tasks, and perhaps for maize to be ground in a powered mill. Here a girl may train as a primary health care worker. Teachers and doctors may be women, but come from outside the community. Generally, a woman's opportunities, whether she is rich or poor, educated or illiterate, expand greatly. In a town, some entre-preneurs create businesses (see p. 102), open restaurants, or become beauticians; work for the poor and unskilled is still poorly paid, but at least it exists.

Where we began: gardens and skills

In the region of Los Tuxtlas, Veracruz (Map 1, p. viii), a cluster of volcanoes is drenched with heavy rain from the Gulf of Mexico. Before 1960, rainforests swept down almost unbroken from mountains to beaches, but now they have been almost stripped by pioneers. We came here in 1990, attracted by the writings of two Mexican botanists, María Elena Lazos Chavero and María Elena Alvarez-Buylla Roces (1988), and an agronomist, José Raul García Barrios (Alvarez-Buylla Roces et al. 1989). They found the 'home gardens' in Balzapote to be the only sustainable land use in the region, in strong contrast to the cattle-raising. These gardens were worked by all the family: could they, we wondered, solve some of the women's problems?

61

In Balzapote, the botanists found the gardens to be spaces for both living and working. The garden (*solar*, kitchen garden, backyard) really includes the house, making a house-and-garden unit for people and animals. It may appear a chaotic, overgrown jungle, but it has a complex structure of a yard (*patio*) for sitting, working, feeding animals or drying clothes, a garden of flowers and low-growing plants (*jardín*) and an orchard of bushes and trees (*huerta*). The mother is generally in charge of the yard and garden, the father of the orchard. The botanists recorded 338 species growing in Balzapote, wild and cultivated: 127 ornamentals, eighty-six for food, thirty-one for medicine, others for seasonings, shade, firewood, glue, building or rituals, while only eighteen seemed not to be used and were indeed weeds. Little is sold and most is for home use.

> The father and the older sons are in charge of acquiring the knowledge involved in the handling and use of cultivated trees. Mother and older children are in charge of obtaining the plants for the garden (mostly ornamental, medicinal and seasoning species), as well as investigating the way of growing and using them. The role played by children is very important, since they introduce to the home garden new useful species [. . .] So, the home garden is a place of agricultural experimentation where all the family take part.
>
> (Lazos Chavero and Alvarez-Buylla Roces 1988: 56)

The system has many attractions. Most plants are perennial, so the soil is little disturbed, and chemical fertilisers and pesticides are hardly needed, because the diversity both feeds and protects the crops. Tools are cheap, inputs are low and most work is by the family, sometimes with a little hired labour. Work and production are year-round. The system is highly sustainable, unlike most forms of production in the Mexican tropics, but is shrinking in the face of cattle-rearing and commercialisation. The original slash-burn agriculture of the *milpa* was complemented by meat, fruits and other things from the garden. Now, as cattle replace the *milpa*, packet teas and medicines are replacing those from the garden. The gardens are dwindling in the face of the cash economy. They could expand if there were markets for their products, but these are not appearing.

The outsiders' solution

This sustainable, labour-intensive system attracted us. Right across the humid tropics, specialists are trying to develop sustainable agroforestry as an alternative to the many unsustainable systems such as ranching. (In agroforestry, crops and/or livestock are produced under timber- or tree-crop trees, making efficient use of labour and land for small farmers (Merchant 1992).) In diversity and productivity, gardens reach their peak in Los Tuxtlas, but even in the thin, dry forests on the limestones of Campeche, the bleak landscape is splashed with bursts of green in the gardens of indigenous pioneers. In Mexico, sustainable agroforestry has been developed by the pioneers themselves. Could it be an answer to the agrarian crisis and particularly to women's search for income-generating opportunities?

A woman's work is very much what she can do in private, domestic space, in her home and garden (Table 2), save for fetching water, which is the work of women and children whether from the yard or kilometres away. Laundry may also take a woman out to a water supply in public space if she has none in her yard (Table 2). The whole *solar*, house and garden, is domestic space and many chores are done outside while dogs, chickens and even pigs roam into many houses. Many women grow medicinal herbs in their gardens, but the knowledge is being lost as patent remedies and powerful drugs (often inappropriate or dangerous) replace them. Hygiene is often not valued highly, but cleanliness is and women work hard to achieve it, sweeping and laundering. Grinding corn and making it into tortillas takes them several hours a day. Most water must be carried and ideally boiled. Most women cook with wood fuel and must inhale the harmful smoke. Most keep poultry and pigs which both make an important contribution to diet and act as small walking banks for emergencies. Women spend most of their lives in their homes and gardens (Table 2, Figure 1) and value this achievement, which none the less limits their options in income-generation.

An important part of our study was to explore women's solutions to the problems they perceived. As we expected, ranked high among their problems was the lack of opportunities for women or men to earn or generate income. Naïvely, we expected them to see a chance to market more goods from the garden, as we were aware of the premium attached to 'organic' products in

Table 2 Mexican surveys: selected tasks by locality; ratio between women and men who 'always do' the task (age 13 and over)

			Area		
Task	Los Tuxtlas (Veracruz)	Isthmus (Oaxaca/Ver)	Palenque (Chiapas)	Balancán (Tabasco)	Escárcega (Campeche)
Cook for labourers	7:1	4:0	4:1	0:0	0:0
Laundry	17:1	25:1	25:1	8:1	20:1
Cooking	32:1	39:0	7:1	9:1	7:1
Hens	6:1	27:1	7:1	7:1	4:1
Fetch water	7:1	14:1	2:1	4:1	2:1
Pigs	3:1	3:1	4:1	5:1	3:1
Fetch wood	1:6	1:5	1:3	1:3	3:1
Garden	1:1	1:1	1:2	1:3	1:2
Milking	1:26	1:16	1:3	1:3	1:2
Agriculture	1:10	1:20	1:11	1:8	1:7
Cattle	1:10	0:7	1:9	1:10	1:5
[n:]	[300]	[85]	[149]	[162]	[156]

Source: Authors' fieldwork
Note: The small samples for the Isthmus exaggerated ratios

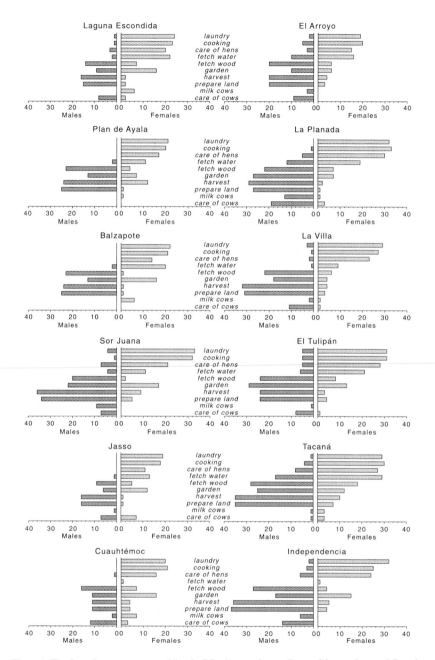

Figure 1 Twelve pioneer communities in Mexico: tasks performed by males and females in the survey households

64

Europe and North America, to 'rainforest ice-cream' and 'forest-friendly' snacks. In Mexico, fresh fruit drinks are a revelation to visitors, but in Europe or North America, canned or frozen mango or passion fruit come from high-input, monocultural plantations, heavily dependent on herbicides and pesticides. To us it seems tragic that these sustainable, labour-intensive systems and their skills may be lost before the market can discover them. Pioneer women are well aware of the marketing difficulties they face, and see no future for their gardens.

FOURTEEN COMMUNITIES IN FIVE STATES OF MEXICO

Los Tuxtlas, Veracruz

It is important to give a brief description of the communities, to display the diversity within communities and the differences between them as well as the continuities. The women quoted in Chapter Five and at length in Chapters Eight to Eleven live in very different places and each experiences her own as a whole, not as the conjuncture of variables which we see. The presence or absence of earth floors, or running water, or electricity (Tables 3–5), or alcohol available to the men are, however, geographical features of importance to all women, as they told us. We shall describe in more detail the four communities from which we are publishing life stories.

All five communities in Los Tuxtlas (Table 3) are within a lush, wet, volcanic area of some 50 square kilometres, but economically and socially they are very different. Sor Juana Inés de la Cruz, for instance, is a prosperous, productive *colonia* of private farms, some of 200 ha. and more, and of desperately poor labourers. The other communities are *ejidos* with land divided equally among members, so that the average plot size determines the prosperity of the community (Table 3), although, again, there are labourers. Cattle now give better returns than cultivation but provide fewer jobs, for one man can care for fifty head of cattle, or milk twenty-five. (Few women care for cattle, and very few milk (Figure 1).) The milk is sold to Nestlé, an important agent in the deforestation of south-east Mexico.

In Plan de Ayala and Laguna Escondida there is too little land to provide a livelihood, so the men must work for landowners and even go away for months to earn a living. Yet there was hunger for this land, and when the pioneers came to Plan de Ayala in 1965 they invaded rainforest belonging to the private landowners of Sor Juana. Carmela (Chapter Eight) faced the guns beside her mother when the landowners of Sor Juana tried to drive them out. All in Plan de Ayala are still poor and live poorly, without water or drains. As in Jasso, El Arroyo and Tacaná (Tables 4 and 5), most people in Plan de Ayala relieve themselves in the bushes. Children and adults are infested with parasites while some have the fatal *chagas* disease, contracted from bed-bugs.

Girls marry young in all the communities, between 13 and 17 years of age.

65

Table 3 Selected characteristics of communities in Los Tuxtlas (State of Veracruz), Mexico

	Laguna	Plan de Ayala	Community Balzapote	La Corregidora	Sor Juana Inés
Interviews	15	22	15	0	23
Life histories	0	4	0	4	4
Workshops	0	0	0	2	0
Status	*ejido*	*ejido*	*ejido*	*ejido*	*colonia*
Date[a]	1960s	1960s	1960s	1940	1936
Inhabitants	300	500	1,000	4,000	3,500
Plot size, ha.	[b]	5	[b]	20	1–200
Forest	some	little	little	none	little
Rainfall	*c.* 4,500 mm	*c.* 5,000 mm	*c.* 4,500 mm	*c.* 4,500 mm	*c.* 4,500 mm
Cattle/ha.	2	2	2	3	3
Polarisation	low	low	high	high	very high
Hours to town[c]	3	2.5	3	2	2.5
Hours to road[d]	2	1.5	2	1	1.5
% of house-holds with:					
'New' houses[e]	0	5	33	–	30
Latrines/WC	5	40	60	–	65
Piped water	0	32	0	–	70
Health post	no	no	yes	yes	no
School[f]	no	no	yes	yes	yes
Women's jobs[g]	none	none[h]	few	many[h]	few[h]

Source: Authors' fieldwork

Notes:　[a] Date of settlement
　　　[b] *Ejido* not formalised, no plot size
　　　[c] Town: seat of municipality
　　　[d] Paved road (time: in hours by truck)
　　　[e] 'New' = brick/cement dwellings
　　　[f] Secondary (all have primary)
　　　[g] For the landless
　　　[h] Apart from picking chillies
　　　　No entry = no data

Many women rarely leave the community or even the garden – in Laguna Escondida, they say that a woman may not go out to visit her mother without permission. As elsewhere, men have bars, football and baseball, while women meet to do the laundry or in church or the Adventist temple. Women dislike the new emphasis on cattle-raising because when there were crops they were stored to be eaten, but now women must ask men for money for food for the house-hold (workshop). The Municipal President has ordered that no alcoholic drinks be sold at weekends, but this edict is generally ignored.

The Isthmus: Veracruz and Oaxaca

Francisco Javier Jasso (Oaxaca) and Cuauhtémoc (Veracruz) (Table 4) lie in the centre of the Isthmus (Townsend with Bain 1993). The land in Jasso was worked collectively until 1989. No alcohol is sold since the women persuaded the *ejido* to ban it, so that men must go out to drink or bring it in. Cuauhtemoc is a fair-sized town, a private colony in Veracruz, dating from the 1950s, where people live mainly by cutting timber, now controlled by the government. Alcohol is sold widely, including illegal, toxic varieties under the counter. Women here (in a town) are much more economically active and some are successful entrepreneurs despite having married very young, for real opportunities exist for women with economic resources.

Table 4 Selected characteristics of communities in the Isthmus (States of Oaxaca and Veracruz) and Palenque (State of Chiapas), Mexico

	Jasso	Cuauhtémoc	El Arroyo	La Planada
		Community		
Interviews	15	15	22	23
Life histories	0	0	2	2
Workshops	0	0	1	0
Status	*ejido*	*colonia*	*ejido*	*ejido*
Date[a]	1938	1950s	1930s	1945
Inhabitants	1,000	9,000	500	3,500
Plot size, ha.		50	15	8
Forest	some	much	little	little
Rainfall	*c.* 3,000 mm	*c.* 3,000 mm	*c.* 2,000 mm	*c.* 2,000 mm
Cattle/ha.			1	1
Polarisation	medium	high	high	medium
Hours to town[c]	2.0	1.5	0.8	0.8
Hours to road[d]	1.5	1.0	0.3	0.3
% of households with:				
'New' houses[e]	0	5	28	35
Latrines/WC	5	40	39	83
Piped water	0	32	0	0
Health post	no	yes	no	yes
School[f]	yes	yes	yes	yes
Women's jobs[g]	few	many	few	few

Source: Authors' fieldwork
Notes: [a] Date of settlement
[b] *Ejido* not formalised, no plot size
[c] Town: seat of municipality
[d] Paved road (time: in hours by truck)
[e] 'New' = brick/cement dwellings
[f] Secondary (all have primary)
[g] For the landless
No entry = no data

Palenque: the Lacandón rainforest, Chiapas

The lowlands north of Palenque, a town at the eastern end of what was the Lacandón rainforest, have been settled since the 1930s. For a rural area in Mexico, their access to urban services is very good (Table 4). La Planada and El Arroyo are adjoining *ejidos*. Unusually, the majority of families have land and at least one cow, but family incomes still depend basically on paid labour. Much of their own land is now in pasture, often rented out to ranchers with the rent spent on consumer goods from televisions to refrigerators. Most cattle are owned by a few, with an average of 100 head per worker (Pontigo Sanchez 1990). The lives of women in these communities contrast with those in Los Tuxtlas, for many girls from families with land complete secondary school, and some may go on to teach or to train in beauty care, in dress-making or as secretaries. The average age of marriage is higher and there are few early pregnancies.

El Arroyo, where Elena lives (Chapter Nine), is deeply divided by long-standing family feuds. A few years after the pioneers arrived here sixty years ago, violent conflicts arose between the Diaz family and the Hernandez family and many were driven out at gunpoint. These two families still dominate and are the most prosperous, controlling much of the land. The feud is still bloody. El Arroyo was a model cattle *ejido* in the 1980s, a leader in organisation and production, but this success collapsed around internal conflicts. The *ejido* has attracted remarkable and largely unsuccessful state aid, to plant oranges (which failed, leaving a large concrete water tank unused), to grow rubber (lost in a drought), to dig fish-ponds (new, but unsuccessful elsewhere). The state Department of Family Integration (DIF), built a Centre of Family Support in 1990 under the national Solidarity programme, an elegant modern building, including concrete basins to wash clothes, showers and public lavatories (kept locked). It offers courses in bread-making, tailoring, sewing and cooking and has a 'library' (without books), a television room and table tennis. Elsewhere, women speak with longing of leisure facilities, but here the feud keeps many away from the centre.

Balancán: plans and deteriorating development in Tabasco

North-west of Palenque, in the state of Tabasco, Plan Balancán–Tenosique was developed, beginning in 1972, as a colonisation project to produce cattle, the prime motive apparently being to 'occupy' land close to the Guatemalan frontier. Fernando Tudela *et al.* (1989) judge it unsuccessful because so many planning goals were not fulfilled for investment, forest protection, irrigation, agroforestry and *chinampas* (rafts with soil on top to grow crops on later, imitations of a successful pre-Conquest technique). The yields of meat and milk are low and some *ejidal* land is still rented to ranchers. The Plan area now shows 'deterioration of infrastructure and equipment: ruined roads, half-flooded villages, blocked drains, fallen power lines, abandoned machinery rusting in the rain' (Tudela *et al.* 1989: 232).

We worked in La Villa in the Plan and in nearby El Tulipán, outside the Plan (Table 5). For nearly twenty years La Villa was run collectively, but it is now progressively being privatised. El Tulipán, on the contrary, has recently and successfully moved to communal production under an Oxfam–Belgium project. It is the most prosperous community in which we worked, farming with industrial inputs, like El Distrito (see p. 39). La Villa, part of Plan Balancán–Tenosique, presents an odd combination of poor urban living and subsistence agriculture (see Chapter Ten). Here we found the most services (Table 4, p. 67), the most bitter complaints of poverty and a community suffering from

Table 5 Selected characteristics of communities near Balancán (State of Tabasco) and Escárcega (State of Campeche), Mexico

	Community			
	La Villa	*El Tulipán*	*Tacaná*	*Independencia*
Interviews	23	22	22	23
Life histories	2	2	2	2
Workshops	1	1	1	1
Status	*ejido*	*ejido*	*ejido*	*ejido*
Date[a]	1938	1950s	1980s	1972
Inhabitants	400	500	600	400
Plot size, ha.	50	[b]	15	[b]
Forest	some	much	some	little
Rainfall	*c.*1,500 mm	*c.*1,500 mm	*c.*1,300 mm	*c.*1,400 mm
Cattle/ha.	1.3	1.0	0.4	0.3
Polarisation	medium	medium	little	high
Hours to town[c]	3	1	4.0	1.08
Hours to road[d]	1	0.3	3.0	0
% of households with: 'New' houses[e]	83	23	0	35
Latrines/WC	91	73	36	52
Piped water	100	5	0	95
Health post	yes	yes	no	yes
School[f]	yes	yes	no	no
Women's jobs[g]	few	some	few	few

Source: Authors' fieldwork

Notes: [a] Date of settlement

[b] Much land in dispute with landowners in El Tulipán; collective land being redistributed in Independencia

[c] Town: seat of municipality

[d] Paved road (time: in hours by truck and, for Tacaná, on foot as well)

[e] 'New' = brick/cement dwellings

[f] Secondary (all have primary)

[g] For the landless

No entry = no data

69

poor planning and neglect. The planners designed a town by bringing four villages together to provide a clinic, schools, electricity (irregular and expensive), water (irregular) and drains (inadequate). The site, however, is subject to frequent floods and filthy water often stands in the streets, and comes into the houses. La Villa was the only evil-smelling village we visited. As in the settlement scheme at Uxpanapa (Ewell and Poleman 1980), the cement houses are too small for the families. Because the houses are closely packed, conflicts often arise betwen neighbours about the children's behaviour. The Plan provided for cooking with gas stoves in the houses, but people have built raised hearths outside with thatched roofs. They allow the pigs and chickens in the house and do not use the lavatory 'because it is too close to the hearth'.

How can rural Mexicans with running water and electricity (albeit irregular), televisions and electric fans describe themselves as poor? Planning has brought 'mod cons', but also very unprofitable and indebted collective work, recompensed at 40,000 pesos per month (US$13 or £8). Some men had been paid nothing for six weeks, so that many women talked of whole days without food, and of a diet of tortillas, beans and maize ground in water, which is inadequate for young children. This is modern poverty, with hunger amid consumer goods, urban services which break down and a desperate shortage of cash. The original site of La Villa, they say, was above the floods, had big gardens and was closer to the ranches for paid work. The Plan moved it here, so women's opinion of the planners is very low (Chapter Ten). There are no bars, but drink is brought in. Television, especially soap operas, is important in women's lives and, as we shall see, talking with other women is not very respectable (p. 117).

In El Tulipán, Oxfam-Belgium has a three-year project for commercial production which places immense emphasis on training in administration and marketing. If the project continues to be successful after Oxfam leave, it may be a model for tropical Mexico.

Campeche: pioneers

The state of Campeche is very different from all these places: remote, dry and one of the last places for recent settlement. The hardships which pioneers tolerate here in search of a possible livelihood (Table 5, p. 69) are a painful indication of conditions in the long-settled areas which they have left. Yet, 'what we miss most here is our families', said women in the workshop in Independencia. People have come from all over Mexico. We shall describe these communities in more detail, as pioneering is more recent, and allows a far more emic, insiders' account. Tacaná is poor and remote (8 km from a drivable track, 22 from a paved road), while both Naranjales and Independencia lie beside a paved road. Because rainfall is low and streams are scarce, the great problem here is water, for crops, animals and people, and in 1990 and 1991 losses were immense. Men's work is still on the land (Table 5). Although a 'day's work' may be from 7 to 11 a.m. because of the heat of the sun, men offer no help with

70

other work. In Tacaná, all water must be carried a kilometre but only a third of men help (Figure 1, p. 64), and then usually with a pack animal.

When Independencia was founded in 1971, the water was 1.5 km away and the women had to carry it all. Food was short and of poor quality, and soap for the laundry was too expensive to use. Problems also arose from the contrasting origins of the settlers. In the 1980s, a state project cleared 320 ha. for the planting of mechanised rice. Loans paid for the seed, equipment, wages and health insurance, yields were high and the productivity was the highest in Campeche, 3 to 4 tonnes per ha. But, say the technical experts, there was too little moisture for herbicides and pesticides to be effective. The project failed, the tractors were sold, the communal land divided and the prosperous families are seeking to change to cattle-raising. The 'time of the rice' made the community rich, and enabled people to buy cattle and trucks, build houses and fences and dig water-holes, but it has become stuck at this stage. Again, the lack of water is a problem and wells are 84 m deep.

Most women travel frequently to Escárcega either to sell goods and shop or to see a doctor. Many have television sets. Women are confident with strangers, open and talkative. Yet they face a desperate shortage of jobs. Women complain, 'We can't keep the family together because there is no work here and they go away to look for work' (workshop). The women want a packing plant where the women can work (workshop). Young, educated people in particular leave, although others, unskilled, still come in search of work. Parents are often much missed, and people from this prosperous village spend a great deal in visiting them: 'We don't like being women because we are not near our parents' (workshop).

Perhaps because of the excellent accessibility and good amenities, the Department for Family Integration (DIF) has sent a series of trainers (*promotoras*) to work with the women in Independencia. These trainers have been extremely important, for instance, in teaching embroidery and weaving which are not traditional activities. Amalia, the latest trainer, was very popular (see p. 119). She set up adult education, secondary as well as primary (see p. 86). She taught women to value themselves as members of a society in which they have rights, not only as mothers, but as thinking, important beings who have their own dignity and their own right to think and feel. To improve nutrition, she taught them how to cook soya, how to chlorinate the water and arranged for some cheap rations. She organised courses on family planning, hygiene and health, and on the prevention of illnesses such as uterine cancer, cholera and AIDS. She obtained birth certificates for people who had none and organised civil marriages for those who wanted them. All these initiatives were extremely well received, but clearly far more time had been given to income-generation and to women's traditional role as caretakers of family health than to questions of gender relations. There had been no discussion of domestic violence or alcoholism, although the women see them as leading problems.

Some women at the workshop in Independencia expressed unusual opinions.

A majority still expressed satisfaction at being women, but some complained of suffering more than men, of having less opportunity, of being unvalued by men, of tedious, boring work, of being trapped with their children, unable to earn. We were told that these ideas did not come from Amalia, so perhaps the urban contact or the television have been more important.

Naranjales was founded in the early 1980s, beside a highway, when three families, who had been driven out of another *ejido* with machine guns, secured plots of 100 ha. for each member of the *ejido*. These families still dominate the other members of the community. We conducted no survey or workshops here, but heard from Mariana (48) and Imelda (55) (quoted in Chapter Five) how, for eight years, all water had had to be fetched, some from another *ejido* 16 km away, 'on foot or by truck'. Often women collected water from ruts in the road to wash clothes and dishes.

Tacaná is the poorest and most remote of the communities in which we worked (see Chapter 11). Founded in 1983, it met opposition from landowners and had six years of struggle, which only ended when some *ejidatarios* journeyed to Mexico City and camped outside the presidential palace until the President saw them. (Appeal to the centre of power is important in Mexico; see p. 51 for the process of *ejido* formation.) Settlers in Tacaná come from Chiapas, Veracruz, Tabasco, Campeche and Mexico City itself. At the beginning, they lived on tortillas, ground maize in water, chickens and wild animals such as armadillos, turkeys, *pacos* (wild rodents), wild pig, pheasant and deer. Wildlife was abundant for people also remember jaguars, monkeys, squirrels, parakeets and rattlesnakes. Malaria has been a serious problem and the state still comes to spray every six months. Tacaná is carefully laid out 10 km from a track (which may be empty of vehicles for days), motorable when it is dry for another 12 km to the road. It is sometimes necessary to carry a sick person the 22 km to the road in a hammock and in the wet season people must cross many creeks with water chest-high. Seventy-three families have left since Tacaná was founded.

This pioneer village has been laid out so that it can develop into a town, with straight, wide grassy 'streets'. Houses are floored with dirt, packed hard, with walls of bamboo or planks and roofs of thatch. Any furniture is made in the village and some houses have only a hammock and a rope over which to hang clothes. For light at night they burn paraffin or carry a torch. Women and children carry all the water more than a kilometre from wells, but rarely boil it.

We brought hammocks and slept in the little wooden hut where the teachers sleep when there. We found the school latrines revolting because the children use the floor rather than the holes.

The men of Tacaná themselves made the track to the unpaved road, with axes and machetes, picks and hoes. For them, the high cost of getting the harvest to market is the great problem of the community. Until recently, they had had to sell cheaply to whatever traders came to El Arrozal, but in 1991 Solidarity lent the *ejido* the money to buy a small truck, still to be paid for.

'When it rains, you can't get the crops out. When it doesn't rain, there are no crops' (Laura). The women, however, see water as the main problem. The men have dug wells, but the only one which does not dry up is on the land of a landowner a kilometre from the village. The queues there are always long, and very few can wash themselves or their clothes at the well. Many women have prolapsed wombs caused by carrying water. For women the first objective is the loan of well-digging equipment by the state, although it is not certain that there is an adequate water source.

This is one community where crops have not yet been replaced by ranching, yet problems abound. Of all the communities we came to know, Tacaná is the least involved with the market and has by far the most garden livestock, yet still very few people are adequately nourished: 'My daughter comes here to see us, but the trouble is, she is used to eating every day' (Ana).

Most men and older children can read, but only half the women. Only one man has any secondary education. In theory, children may complete primary education at the school, but the teachers are rarely in the village. It is a bitter complaint in the village that the four teachers do not live in Tacaná but come to give classes two or three days a week and then stay away for perhaps a fortnight, claiming that they have to report on their work. Children in the third and fourth grades can barely read. In their spare time, the men have built three classrooms roofed with thatch, open at the sides and furnished with blackboards and home-made desks. Some 160 pupils are enrolled, but the teachers are often missing.

It seems that in Tacaná there is really no alcoholism and public safety is very good. To us, women in Tacaná seem oppressed and submissive, introverted and expressing their thoughts with difficulty, especially alone in their homes. They nevertheless told us of the needs and wants of the community, taking these to be their proper concern. Here, religion has an important role in supporting patriarchy, for birth control is not accepted. It was here, above all the communities, that women told us that they did not know how to avoid conception. Amparo says that she has enough children and knows how to stop, but is not allowed to because her husband wants 'five more, and all boys for the work'. Most girls marry between 12 and 15. Women describe their satisfaction as being in their children, who occupy most of their time. Most women, particularly the older ones, rarely leave the *ejido* because of the difficulties of the journey. Women meet in the places of worship and at the wells, but claim to make few friends because they come from such different backgrounds.

The men called a workshop for us, in the school, of all the village women; perhaps 150 came. Once together, women were much less meek and presented themselves not as needy and passive but as mothers who know how to make the best of themselves. They were vocal on the problems of water, access, the school and the health centre and called for a mill to grind the maize, and training in dress-making or weaving. Domestic violence they see as a hidden problem. The workshop ended with the women deciding to choose a secretary to write to the Governor of Campeche to ask for a well, with all signing the letter.

LOSING THE FOREST?

When the pioneers have created a landscape of fields and villages and the rainforests are a memory, how do they feel? To the Western media, this is destruction which damages soils, chokes rivers, threatens local climates and may play a part in global warming. To Mexicans and to us:

> The North wants the South to reduce its growth of population and its economic growth to conserve biodiversity and the sumps for the greenhouse gases which they [the North] produce; the South wants the North to pay the cost of its high consumption of energy and of the other natural resources of the planet and of its production of poisonous gases.
>
> (Arizpe *et al.* 1993)

How do women settlers react to the loss of the forests? Those who came as pioneers into the forest reel off great lists of the animals there used to be, whether in the rainforests of Los Tuxtlas or the dry forests of Campeche. But do they all agree with Cristina (65, Cuauhtemoc): 'If we didn't fell the forest, what would there be to eat?'

Internationally, ecofeminism is very diverse. Some ecofeminists such as Vandana Shiva (1989) believe that women are closer to nature than men and that they are more aware of the needs of nature, of the needs of their children and grandchildren and the links between these needs. In 1990–1, Lourdes Arizpe, Fernanda Paz and Margarita Velásquez (1993) asked for the opinions of more than 400 people, from Palenque to Marqués de Comillas in what used to be the Lacandón rainforest, about environmental change. Their findings do not support Vandana Shiva's thinking, for the differences were primarily between old and new communities, Spanish- and indigenous-speakers or ranchers and pioneers, not between men and women.

Jennie Bain, a member of our team, set out to demonstrate that Mexican rural women are more environmentally aware than men. She showed their close involvement with environmental processes but found no examples in the Mexican literature of women engaging in more sustainable environmental practices than men. She still believes women are more aware, but that the evidence has not been found to prove it (Bain 1992, 1993). In 1991, during our household survey, we asked questions for her in seventy-seven households about attitudes to forest clearance. Most of the answers came from women, but we did not record the gender of all respondents. Of the seventy-seven people asked whether they were happy about the forest being cleared, sixteen had no opinion. Seven were pleased that the forest had been cleared, in order to produce food, because it had sheltered dangerous animals, insects and snakes or because the trees bring thunderbolts (compare Arizpe *et al.* 1993: 101); nine others were 'not worried', and forty-five said they were worried.

Our findings do not conform to Vandana Shiva's (1989), perhaps because in Mexico the gender division of labour puts women in less intimate contact with the natural environment. We asked whether men, women or everyone should protect the forest. Of the forty-five worried about the loss of forest, thirty-eight thought the forest had to be protected, including twenty-two who thought it was everyone's responsibility. Ten thought it was up to men (as men cut it down and 'as it is the man who maintains the house') and only six thought it was up to women because, they said, they are 'more aware'. However, a small number of women in very different and distant places do seem to conform to Shiva's model:

> As a woman, you think about the welfare of your children, and if they cut down all the trees, they'll leave all the forest destroyed and a desert [. . .] The woman wants to see a better future for all of us, so that we may breathe clean air.
>
> (Catalina, Independencia, Campeche)

> Women have to watch out that the men don't use up all the forest.
>
> (Carmela, La Planada, Chiapas)

We did not explore whether the men are more or less worried than women about the loss of the forest. Manuel (El Arroyo) is very critical of clearing:

> We came to kill the trees [. . .] The forests had no undergrowth and were very beautiful, the streams ran clear [. . .] The land was rich [. . .] The atmosphere is so bad now that sometimes it doesn't rain, because the water goes with the vegetation [. . .] The soil is completely destroyed by fire.

Women are diverse everywhere. Here, some see sympathy with animals and trees as a feminine thing, but Isabel (La Corregidora) has very different feelings,

> We saw jaguars, those things – big rodents – armadillos, racoons, badgers – then wild boars, and in the streams, shrimps. And we were scared because we didn't know that kind of jaguar, the pumas there used to be, pheasants [. . .] But then afterwards we took control of the forest and went hunting by night or in the day. I had an eighteen-shot rifle – I loved shooting at the animals, killing pigeons, doves. Once I killed a small jaguar of about, mmm, 40 kilos. It was in a tree. The chickens panicked, and I went – but then you had to be careful underfoot because of the snakes, because there were lots of bushmasters. I don't know how many of those I killed because there were so many. I killed them with the rifle, with the machete – little ones, but they still count, don't they?

The forty-five who told us they were worried about losing the forest may have said that because they felt that we wanted them to, but the detail of their answers is interesting. They said that when the forest ran out

- there would not be enough rain (sixteen people);
- there would be no land to work (fifteen);
- the animals would be lost (and with them a source of food) (twelve);

- there would be a lack of oxygen or clean air (ten);
- there would not be enough to eat (eight);
- there would not be enough firewood (five);
- there would be no wood to build houses (five); and
- the beauty would be lost (five).

Other anxieties were the loss of shade, knowledge and fine timber, and possible increases in disease and drunkenness. In local terms, all are women's issues although some are also men's.

Our biggest clusters of 'worried' people were in Campeche, where pioneering is more recent and some forest survives. Here, more people have an opinion and more are disturbed about the loss of forest than elsewhere, although the cause of the concern was very different in the two communities. In Independencia, people worried most about future air and water, and we think both television and urban exposure may be important in stimulating this concern, for Independencia lies by the highway and has many television sets. Worries about 'clean air' and 'oxygen' here and in the Lacandón rainforest may come not from ideas of global warming but from television programmes about the pollution of the very different environment of Mexico City by vehicles and industry (Arizpe *et al.* 1993). In Tacaná, which has no electricity and no television and is a long walk from anywhere which has, people worry most about whether there will be land to work when the forest is gone and women see this as central to their children's futures.

Rain tops the list of fears. In both 1990 and 1991 we were told in all communities how much less rain there is than when the communities were founded (in 1991 there was a severe drought and people talked a great deal about the loss of rain). Only one person, Jesús, in Tacaná, told us that the drought was not caused by clearing the forest. Contrary to the experience of Lourdes Arizpe's team (1993: 126), we were told by many that 'the trees bring the water'. María de Jesús (85, Sor Juana) spoke for most people:

> Well, now we're all convinced by what they say, that the day we felled the trees will be a shame to us [. . .] Many have even wept with this weather now that it hasn't rained. And it was all because of this, they cut down many trees, and the forest attracts the water, so it was their fault.

Laguna Escondida has disputes with the National Autonomous University of Mexico about the boundaries of a biological reserve, but even there people pay lip-service to the need to protect the reserve 'to keep the rain coming'. Quite a few pioneers seem to have heard ideas propagated by the North, whether from the media or from school. There is an awareness which ecofeminists wishing to protect the forests could develop.

Opinion is divided as to whether men or women know more about plants in the gardens, but generally we agree with Olga (Laguna Escondida): 'Some men know a lot, but for most of it, it's the women.' Some men are fascinated by plants, trees and medicines. As it is men who go to the forest, generally go out much more, travel more widely and meet more people, men have more opportunity to develop new and more specialised knowledge.

CONCLUSIONS

Women's lives are transformed again and again as the relationship with the environment changes in new communities in the *selva*. At first, life is hard and children are at a premium, but later, although more services may develop, jobs become desperately short and education lacking. Women ask for training, a chance to earn and opportunities to spend more time with other women to confront personal problems. Unfortunately, there seems little opportunity to expand the sophisticated, labour-intensive agroforestry which families have developed in their gardens and in which women have considerable expertise, for the necessary markets are not developing. Thus, women's economic options are frequently limited by their wish to stay in the house and garden. If a link could be made between the multiple products of the gardens and the world market for tropical goods, it would provide a very attractive partial solution to the women's problems.

This is our view as outsiders, an analysis, an etic account of the condition of pioneer women. As we see it, pioneering separates women from their network of support and protection at the very time when their men are living very stressful lives, so that the isolation of women may contribute to wife-beating and male drunkenness. Then, as the forest is exhausted, many men become unable to feed their families and violence and drinking are perhaps increased. Future studies in actively pioneering communities may be better able to establish how gender relations change in Mexico during and after pioneering.

Part 2

OUTSIDERS AND INSIDERS

5

MEXICAN WOMEN PIONEERS
TELL THEIR STORIES

WOMEN'S WORDS

The settlements we studied in Mexico differ widely in the resources available to them, in degree of accessibility, in past history and in present economy. Yet when women of different ages who had moved at different times to different places talked about their experiences of colonisation, we were surprised again and again by the strong similarities in their tales. We want in this chapter to present some of women's experiences in their own words. Since we have extracted their words from interviews and life stories, this is still an etic, outsiders' view, selected and organised by us. We believe that it still offers insights into the lives of pioneer women.

Methods

To write this chapter, Janet Townsend first read all the life histories straight through, noting themes which seemed to be discussed by many women at length and with interest. She then extracted comments on these themes and grouped them together, before adding comments on the same themes from our notes on unstructured interviews. The chapter was thus written around the women's words.

COLONISATION

Why colonise?

A succession of government policies, beginning in 1940 but much more effective from the late 1960s (see p. 51), encouraged people to settle on 'new' lands in order to reduce the pressure in other areas for more land reform (Arizpe *et al.* 1993). A few, like Angela (p. 84), who went to private *colonias*, had some resources but the new *ejidos* were primarily the refuge of the destitute. Most women who told us of their lives came because they 'had no land' and were ready to do any work to get some (see Chapter Eleven, p. 184). Susana (aged 68,

Sor Juana) is very definite: 'Because of poverty. Because of our poverty. Poverty brought us here. Where we were, we were poor, poor, poor.' Lourdes (48), her husband and one child came from highland Chiapas to Tabasco 'to see if we could find a future for the children'. They travelled, working as caretaker and cook on ranches, for twenty years, 'earning miseries' and raising eight sons before they acquired their first piece of land. However, from there they still moved to Naranjales. Most of the twenty-six women who told us their life stories had poor, rural backgrounds; only four were born in towns. Some had been driven out by force from their old homes. Ernestina (38, Independencia), for example, had been dispossessed twice. First, when she was still a child, the whole family left the distant state of Guerrero after her father was murdered in the 1960s. Later, they joined a land invasion in Oaxaca and suffered an experience on which Ernestina blames a stillbirth:

> There, in that *ejido*, we were resisting some landowners, rich men who wanted to take the land, because the rich don't let ordinary people own things. And they [the landowners] sent for the federal police, they sent them in during the night and I was terrified – they came round waking everyone and beating them up. They took the men to a big mango tree which was there [. . .] tied them with a lassoo here [she mimes round the neck], pulled them up and beat them.

Imelda (55, Naranjales) suffered under similar circumstances in Campeche in the 1980s:

> They came and threatened us with machine guns – they all came armed and sometimes it was just us women there in the houses and they came and threatened us. We'd had enough of it – we were afraid . . . In the end, when we'd left, many families still stayed, and they came and beat up the men and burnt their houses.

Some settlers had been dislodged by family conflicts, some by losing everything. The husband of María del Jesús (85, Sor Juana) spent all they had in betting on horse-races and cock-fights: 'Then I said, "And now are you ashamed of yourself? If you are going off to work on the land, look, I'm thinking that they say there's a lot of land in Sor Juana . . . "' Often, the woman did not want to migrate, but felt she had to, like Florencia (77) in Sor Juana:

> He came to look for land to work. I've told you we had no land to work . . . I said to my mother, 'Mum, he's going to take me to Sor Juana.' 'Well, girl, you're married now and with small children: you can't say no, where he goes you have to go.'

Gloria (47, El Tulipán) is younger, but 'He had to go here, there and every-where, looking for work and I had no choice but to follow along behind him, because he is my husband and I have to follow him.' Some had heard positive accounts of the place from friends and many followed a relative. Antonia (47) came 1,300 km to Independencia to join her parents when she lost her home in Mexico City to an extension of the underground railway. Victorina (33, Independencia) feels that she had been married and brought to the forest just to provide a companion:

> Well, he, well, he came in February and it was in July that he asked me to come here, but I'm telling you that later I found out he'd had three girlfriends [she laughs] in that time, he'd had three girlfriends [. . .] He went to Apaseo and asked a girl to come and she wouldn't. She already had a better boyfriend. He went to Tinaja [. . .] and asked another girl to come and she wouldn't, because she'd have had to leave her parents and go far away [. . .] and as I was having a bad time, well, my aunts advised me, I decided and I fell for it [she laughs]. I tell him now it wasn't love [. . .] He says, 'If you hadn't come, I'd have brought some woman somehow.'

Hardships

Most of the women take great pride in the hardships experienced and overcome. Obviously, it was harder for the women from towns, and above all for those with small children, even as late as the 1970s (Chapter Nine, pp. 161–2). Most see it as very important to keep small children out of the forest, unlike the Lacandón Indians whose children are competent in the forest at the age of 5. Guillermina (66) came to La Corregidora in the 1940s when there were only reed shacks:

> There were just seven houses. And just paths – you went everywhere by path, there were no roads. No drinking-water. No electricity. No communications, so with a sick child – if there happened to be someone around skilled with herbs, you'd ask them to cure it, that was fine, you see. If not, just imagine what it was like getting to a town, on a mule or on foot . . .

Ernestina (38) was not pleased by Independencia in the 1980s:

> There was no water here in the *ejido*, we had to go to the pond, to the puddle, to get water. And the children just filled up with bites and spots, that was horrid – there are these things called *colmoyotes* which get into the flesh, into the skin [. . .] and that *chicle* fly [chicle is the raw material for chewing gum] which eats up people's ears. One fly bit my husband [. . .] bits [of his ear] came off [*sic*], and there's no cure, because there's just no medicine for it.

Independencia is now on a highway and services have come quickly, but many other remote communities still lack all services after decades. The wretchedness of diseases and deaths in the early stages of all communities are naturally recounted at great length, although the continuing intestinal and respiratory infections are rarely mentioned.

The house is very important, as shelter and symbol. Petra (60) came to La Villa in the 1960s and is still proud that 'When the baby was born, he was building me a house so people wouldn't see me. Yes, he really managed well so that I couldn't be seen and the wind couldn't get in.' There are sagas about the building and financing of huts, houses, schools and sometimes of the wooden houses burning down and people losing everything.

Transport has always been a big issue. In Tacaná it is still a 9 km walk to the motorable track. Where access is poor, life revolves around provisioning and access to medical care. Isabel (51) came to La Corregidora twenty-five years ago:

> Every week you had to go [to the town] to shop for the week, beans, rice, oil, soap, salt, sugar – you had to buy them all for the whole week because there was nowhere here. Now, if I run out of sugar it isn't far to the shop – now there are lots of shops it's all easier, and it is in emergencies too – I just go to the road and any vehicle will take me.

Food is another central topic – poverty, hunger, the difficulty of getting provisions, the variety of wild foods from the forest, the fate of the crops on which survival depended. Animals from the forest were 'good' food, but Isabel (51, La Corregidora) refers to plant food as 'weeds', on a par with snails.

In an emergency, money is a desperate problem. Celia (La Planada) had to beg to get her sick child to hospital:

> I pulled myself together and went to the Town Hall and spoke to the Mayor, and he told me, 'Look, nothing can be done quickly [. . .] so I'm going to give you a paper, and you go from house to house to ask for help [. . .]' Oh my God! I went to the houses, feeling as if I was hanging by the ears. What a humiliation! What shame I felt knocking on doors, arriving with this paper to beg for charity.

Pioneer work

Every woman we interviewed talked about the work. Women who did not share the agricultural work often took up selling or, if they could, bought a powered mill to grind maize commercially for the staple tortillas, in order to generate income. All worked hard to build up a yard of poultry and pigs and a great deal depended on the home garden. Many tasks were new: 'Here I work more, because, look, in a town you buy tortillas, they sell *pozol* [maize ground up in sugared water] – here you have to grind the maize, put it in the handmill and do it yourself' (Gloria, 47, El Tulipán).

Experience varies according to class. Angela (76, Sor Juana) came from the town, with a more prosperous husband, to private, not state colonisation. She never worked in the forest or fields, but wealth created other work:

> When we got here there was nothing, just forest, so what they did was to bring workers in to clear a patch and be able to build a house. They made it

out of poles and leaves, open all the way round, just roofed to cover us from sun and rain, and there we slept somehow [. . .] We women made the food for the men who came to work [. . .] We had a mill to grind the maize, because too many came to work and had to be fed. Frankly, I never ground maize by hand, because where I grew up my mother sent the maize to the mill, so I never learned [. . .] I looked after eighteen workers, and cooked for all of them with enormous stews in great big pans [. . .] I had chickens, I always had about 200 chickens, apart from the other poultry, laying hens and so on. It was a lot. And I'd kill one two or three times a day to feed the men.

Advice

What do women think matters? We asked individuals and groups what mattered to them, and for their advice to new women settlers. The emphasis is always on work, on co-operation, on unity and on good relations with one's spouse, as in Guadalupe's life history (p. 192) or Julia's (34, Plan de Ayala):

It is going to be extremely hard, but those who truly know how to work don't find life so hard. I'm telling you – if the two of you really co-operate you don't go short, but if only the man goes to work and the woman just waits in the house, you don't get anything that way, you don't get anywhere.

Adolfina (44) is president of the painfully divided *ejido* of El Arroyo:

Well, they should behave well with others, and build up networks. Because sometimes, if you don't have such a friendship, or such an acquaintance, how can I put it – you have to live with other people. Otherwise, you can't colonise, if there are family problems or fights between neighbours or between families.

CHILDHOOD

Education

For generations, Mexican girls were confined to the home after puberty and therefore deprived of any formal schooling that was available. This denied them not only education but also opportunities to move beyond the confines of the home, thus further restricting their development and notions of achievement. Sarah LeVine and Clara Sunderland Correa (1993) studied the impact of the massive expansion of educational opportunities in the 1970s on women in Los Robles, in the city of Cuernavaca, and see the social experience of schooling in very positive terms: 'The longer she stays in school, the more likely she is to be convinced of her own efficacy – vis-a-vis males in particular – a conviction that stays with her into adulthood, marriage, and motherhood' (LeVine in collaboration with Sunderland Correa, 1993: 197). In the urban areas, the great majority of young mothers have at least some secondary schooling, but not in our remote rural areas. In our survey, 28 per cent of females aged 12 or above were illiterate, as against 14 per cent of males, and only 17 per cent of females and 22 per cent of males had completed primary education. Of the twenty-six women who told us about their lives, only two had completed primary education and many could not read. Most older women had never attended school (often none was available), but more commonly women had gone to school for a year or so but had not even learned to read. Some had played truant, but more often they had been needed in the home, especially if their mother died, and so could not attend school.

Propriety was, and still is, important and a barrier to girls' education as adolescent girls were and are seen to be at risk away from parental supervision. Petra (60, La Villa) was one of five daughters, none of whom attended school as it was 5 km away and their parents saw the distance of the journey as a great danger to girls. Today, no girls from Laguna Escondida attend the secondary school at Sor Juana, 3 km away, although one does live with relatives in town during term to attend school. (All the communities have full primary schools, so that children needing schooling away from home are older than in Colombia and women are more willing to entrust them to relatives.) Angela (76, Sor Juana) loved school and wanted to be a teacher,

> but my father didn't like that job, because he said I'd have to go out alone, and he was against letting me go out alone, 'No', he said, 'there's a lot of wickedness, and no, I don't want you going out alone. Your mother can't go with you because she has to look after us.'

Where there has recently been the opportunity (in Independencia and La Villa), women have learned to read as adults, and some, like Clara (p. 174) have gone on to secondary education. Most of their children can read (particularly those of the younger mothers). Few can afford to have their children in secondary school, but some have achieved it and even some further education.

In Cuernavaca, mothers see the acquisition of a trade or profession as a young person's most important goal (LeVine in collaboration with Sunderland Correa 1993). In these communities, opinions on the benefits of education are more divided. Some, indeed, set great store by education, but many see the costs of equipment and the loss of earnings as prohibitive. Gloria (47, El Tulipán) says that the school begged her to 'make the sacrifices' for her daughter to proceed into the third year of secondary school, but 'they wanted so much'. There are disputes around this question of economic capacity says Tomasa (42) of La Planada:

> I tell you, they deprive their children of school. There are many families which could do it, but don't let them – for example, my aunt has a daughter who wants to study, and do you think they'd let her? They didn't let her,

because they say that girls will only look for boyfriends at school [. . .] That's what I call selfishness, because they have resources. I'm going to show them through my children – I'm getting them ahead even though I'm poor, I'm giving them the opportunity. I tell you, all mine are at school. The only one who isn't is that one in the hammock [a baby of 3 months].

One daughter of Lucia (38, Jasso) won two competitions in the towns:

The teachers told me, 'She must keep on studying, she's got a head for learning.' But we didn't have enough to send her out there, out of here. If I had had, she would have studied. She might have been able to help us eventually [. . .] There are a lot of bright kids whose intelligence is just wasted [. . .] Men who don't have education work in the fields. That's the one thing they are good for.

The importance of further education is a constant refrain. To obtain it, children must leave home, and their parents must incur the cost of their books, the loss of their labour and the personal loss of their presence, but many women who can afford it make these sacrifices.

Girls' productive work

Most of the women had worked hard as children. The older women and those from indigenous families were more likely to have worked in the fields, as well as in the home. Girls would carry in the products, plant and harvest but, for girls or women, this is always 'helping', not 'working'.

Did you start as a girl to go to work in the fields?

From very young, from the age of 10. It was heavy work with a machete, and planting beans. We went just – because we couldn't manage the axe – just to plant beans or to prepare the maize for harvest, to cut the weeds under the maize, anything we could help with.

(Imelda, 55, Naranjales)

Isabel (51, La Corregidora) sees agricultural work positively:

My brothers worked, and I helped them plant beans, maize, rice [. . .] I wasn't one to lie in bed until 5 in the morning – by 3 I was up, grinding maize, making tortillas, making coffee [. . .] [Guns] were my joy [. . .] I went to the *milpas*, and when I came back there was lots of *masape*, like venison – I came back carrying my animal, because a shot I fired didn't miss [. . .] That was my childhood [. . .] I was a real tomboy. I liked the work in the fields best, to yoke the oxen, cut down the scrub – what I didn't like was digging. Horses were my delight, to ride on horseback. No horse got away from me.

Girls' domestic work

Some women had worked in childhood to generate income, but all had worked hard in the home, learning all the tasks (see Chapter Nine, p. 156). Growing up in big, extended families, many girls seemed closer to and learned more from their grandmothers than their mothers, as in Cuernavaca (LeVine in collaboration with Sunderland Correa 1993):

And then, when you were little, 8 or 9, didn't you learn housework?

Only the washing. Yes, my mother set me to grind, but very little. From time to time, she taught me. But to wash clothes, yes, and I liked it. Because she said to me, 'Look after the baby!', and because I didn't look after it, I had to cuddle it and it cried and wouldn't be quiet. I didn't like it. So, I ran off and grabbed the baby's nappies and went to wash them, and my sister took over looking after the screaming baby [. . .] It was my grandmother who taught me. She taught me everything, how to cook a meal, how much salt, how to make a tortilla, to grind dumplings between the stones. That's what she taught me. And to wash clothes, but I knew how to do that. My mother took us to wash, to bathe ourselves, to comb our hair, to make plaits. And so it was, until I grew up and lived with my husband.

(Susana, 68, Sor Juana)

Puberty

Human bodies, male or female, used not to be discussed in Mexican families, and girls were kept in extreme ignorance. Since 1974, sex education has been mandatory in Mexican schools from the beginning of the sixth grade, but the textbooks are not very informative and, as we have seen, few people in our research areas reached the sixth grade. Most women of all ages told us how ignorant and afraid they had been at puberty, when they had their first period. This was the theme on which the largest number of women talked to us very freely (see Chapter Nine, p. 156, Chapter Ten, p. 168, Chapter Eleven, p. 181). Most were shocked and frightened, some terrified. If they had sisters, they would then be laughed at for being afraid. Four, including one elderly woman, had heard about menstruation from friends and made little of it, but many, not attending school, had no access to other girls unless they had older sisters. Often the tale was a saga, with their first period occurring on a journey or during a crisis, with no one to go to, or too much shame or fear to tell anyone. Some, but not all, had told their own daughters in advance while many claim to leave it to the school. Embarrassment still rules.

Only a few told of herbs to help:

[Victorina's grandmother told her] 'when you're like this, don't carry heavy things', and things like that, nothing else. And what she made for me were home remedies because I tell you the pain made me cry. I – I have writhed on the floor from the agony that hit me. And she boiled up herbs like the flowers of cress, like *ruda* in chocolate and all those things, and this was my medicine, but she never took me for advice.

Ruda, grown in home gardens, is widely used for period pain and is an abortifacient. Other physical advice would be not to bathe, and perhaps not to eat 'cold' things, according to folk beliefs deriving via the Spaniards from ancient Greece (Currier 1966):

> I was frightened, I was frightened when my first period came. I didn't know [. . .] [my mother] said not to eat cold things, nor bitter, nor lemon, 'Because that's bad', she said, 'if not, it will give you a fever'. It never did, perhaps because we were working with the blood running, we worked a lot in the sun, in the forest.
>
> (Petra, 60, La Villa)

Seclusion

It was at the time of puberty that many – but by no means all – had been told how to conduct themselves as women: they needed to 'take care of themselves', not to trust men, not to be alone with a lover. They were no longer girls, but virgins, *señoritas*, to be protected, to avoid the physical and social dangers of public space. They had no idea why. Girls are seen as endangered by innocence and physical weakness and as, ideally, in need of seclusion, impracticable as this may be. All men and boys outside the family become a danger, and virgins must be protected from them by the family.

Most girls past puberty are still much restricted in their contact with men and boys, meeting unrelated males only under strict chaperonage and losing their freedom in public space. Of the women who told us of their lives, only Antonia (47, Independencia, who grew up in Mexico City) became pregnant before leaving home; normally, pregnancy follows elopement or marriage. Ernestina (38, Independencia) was repeatedly beaten by her brothers for being seen with the man she was to marry without her parents' approval.

In Los Robles, Cuernavaca, restrictions have changed greatly over the generations (LeVine in collaboration with Sunderland Correa 1993): before the 1970s, girls would be taken out of school before puberty and would meet no males other than brothers and cousins, while, in the 1970s, they were carefully chaperoned, but, in the 1980s, most girls between the ages of 15 and 19 were still in school, and might be allowed to go out with boyfriends under strict rules. In the communities we surveyed, there has been less change although some middle-aged women had attended dances and a few had had several boyfriends. Girls of 15 (unlike those in Cuernavaca) have still almost all left school and are still usually secluded and chaperoned until they elope or marry. Most women of all ages see their father as the main source of restrictions.

Among the older generation, many had not even had the freedom permitted by chaperonage. As girls, they had simply not mixed with men to whom they were not closely related. In a town, they might have had clandestine correspondence, but here most were illiterate. Florencia (77, Sor Juana) speaks of her married life as happy. Yet she had never met her husband before the day she agreed to marry him – 'almost by force', as her parents required her just to say yes or no. No youth who came to the house spoke to the girls, 'because my father was very proper'. She had previously refused to marry a young man whom she had seen drinking with her father, saying that she did not want a drunken husband like her father, 'No, I believe the world is bigger [than being married to a drunkard].' Some older women are nostalgic for the old ways, as Angela (76, Sor Juana) tells her granddaughters,

> The way a girl dresses honestly disgusts me, because my father taught me to dress normally, you see, modestly, and now I see this, and on television you see the skirts up to here [indicates] and that's why there's so much wickedness now, because the women provoke the men, that's certain.

The consequences of breaches of the rules were severe. Julia (44, El Tulipán) tells of being forced to marry the man in charge of the ranch next door because they were caught talking together, 'looking for the calves':

> He [her father] said, 'No, now you've got to marry because I won't allow this' [. . .] Only because we were talking, because my father had high principles [. . .] From the time we married, he treated me badly, he hit me. When, when I came to see it, I saw him going by, going by with his, with his mistresses. Or he would say, 'Get my clothes ready. I'm going out.' He'd go in the evenings to the town where he had his mistress, there he'd pass the evening happy.

Guillermina (66, La Corregidora) speaks bitterly of her sufferings in married life, and blames her choice of husband and her subsequent troubles on the seclusion and ignorance imposed by her father. She now reads popular accounts of ideal sexuality in marriage and has a new perspective on her life. She had a sheltered childhood and speaks acidly of early marriage:

> A wholly barbarous life, yes, barbarous, that's what uncouth men are like [. . .] treating you like rubbish. They don't want any liberation for women [. . .] I, well, my honeymoon was of ice. I didn't know what a honeymoon was nor what that was nor – no, oh no!, horrible things that happen to you through ignorance – they don't prepare you before [. . .] I married out of boredom with my father, because he didn't take us out [. . .] so I fell for my husband and then truly that was a disaster – but that's what I wanted, and that's what I got.

The rationale for seclusion of daughters in Latin America has been widely discussed and often linked to the Spanish heritage (Pitt-Rivers 1971; Stycos

1958; Pescatello 1976; Douglas 1984; Taggert 1992; LeVine in collaboration with Sunderland Correa 1993). Everywhere in Latin America the practice of seclusion has been breaking down, usually painfully. Girls are still taught to see men as a great danger, although rarely given a reason. Traditionally, a man must be able to guarantee the purity of his women, i.e., the virginity of his daughters before marriage and the fidelity of his wife (Pitt-Rivers 1971; Douglas 1984). Sexuality and talk of it are normally still seen as a threat to society, so that mothers, poor and prosperous, find it very difficult to talk to their daughters.

Physical violence

It is, of course, within the family that girls are likely to be battered. Beating was and is commonplace (Chapter Nine p. 156), and lesser uses of physical correction are seen as natural and essential parts of child-rearing. Several spoke of battering in childhood and scars were shown with pride. We did not find any geographical or historical pattern in child abuse, but of course there were only twenty-six interviews. Mistakes made by children in selling goods were the cause of several incidents of abuse. For instance, Ernestina (38, Independencia) is scarred on face and legs:

> My mother, well, used to beat us, yes . . . But it served us well, because now – we know what it is to run a home and do the work and everything . . . She beat my brothers too, and so they are real men, they work in the fields and know how to do everything, because she taught them. Because as she was treated as a child, so she treated us [. . .] Once she beat me very badly, because I sold 2 kilos of maize [. . .] So she said, 'For crying out loud! Six years old, and you're really good at destroying what's in the house!' [. . .] She tied me up with a piece of flex and she gave me a – she hit me in the face, she laid my face open, every blow broke the skin, floods of blood, and she hit us horribly.

> We were terrified of my father and mother. When a plate broke, they'd always give me a whack on the head [. . .] And I was screaming to get outside, because you see I saw she was picking up a pole. I went running, because – and she caught me. Once she drew blood [tears] because she wanted to get control of me [tears]. My mother threw a pole at me, she hit me with a pole, I bled all down here. I had a bad time. When a plate broke, they hit me a lot.
>
> (Florencia, 77, Sor Juana)

> My mother used to hit us often. My father always interrupted 'Why do you hit them so much?'
>
> 'Because', she said, 'you don't have to be with them all day'.

What did she beat you with?

With a riding whip my mother had, that kind, that kind for a horse which have two leather lashes, that kind of riding whip. She hit us with those, she laid it on thick.

(Julia, 44, El Tulipán; see also Chapter Nine, p. 155)

Most of the women who had been beaten treat their children in the same way, but Julia (44, El Tulipán) claims to do better. Lilia (41), a single mother in La Corregidora, could not find time for her children, yet beat them. Both left her:

I worked, and I got tired, and when I got home I rested, and my children had homework. And I left them alone doing it with their friends, so I couldn't talk to them. And at night, well, I lay down to sleep. So clearly I didn't have time to talk to them, when they came in, I went out to clean the school where they teach secondary with television. And in the end, when they left the secondary school, they went. So – what time did I have to talk with them?

Did you hit your children?

Yes, yes, I hit them, because sometimes they didn't pay attention. I said to the girl, 'Wash the dishes for me', and she didn't wash them for me, or 'Fry me the beans' or 'Boil them', and sometimes they dried up and they were all we had. Well, how did it go with my nerves then? Yes, I had no peace, I was always worried and what I did was to give them their thrashings or slappings. At the end, I could hardly speak to them.

Children often suffered badly at the hands of step-parents and other relatives. Victorina's stepmother

gave the door a push, and I fell down. An uncle of mine was working near there [. . .] and he jumped up and grabbed me and carried me in to the house in his arms because I was unconscious. He took me into the house and scolded my father, but my father never believed what was said to him [. . .] I told him what she did, see, and he said, 'Look, girl, I'm fed up with, with what you tell me', he said, 'you only do it to make me unhappy, to see me quarrel with Lola'. And he [my father] got the rope and threw it over the beam of the house, see, and he held me screaming and put the rope round my neck [. . .] and my uncle jumped up and took the rope off, and told him off [. . .] and said, 'Ask me if what Victorina says is true or not: I was here.'

(Victorina, 33, Independencia)

Several women told of an unhappy childhood with a stepmother. No one spoke of sexual abuse by father or mother, but one spoke of abuse by a stepfather and another wept and was silent. Irma (34, La Corregidora) was

92

abused at 13, and ran away rather than tell her mother. All stepfathers are suspect:

> Now, the girl who just came in is my granddaughter. She has had a stepfather since she was one and a half. And what I advise her is, not to trust him. How many cases have there been that the man abuses his stepdaughter? 'No', I say, 'child, respect the man so long as he respects you. Never trust him, because afterwards he'll go on to other things.'
>
> (Guillermina, 66, La Corregidora)

Unlike menstruation, child sexual abuse is not a subject about which women talk freely. They only weep freely.

Conclusion

Most women have some happy memory of childhood, often of playing with brothers and sisters. As in Cuernavaca (LeVine in collaboration with Sunderland Correa 1993), few were allowed playmates outside the family, and many received their main adult affection and instruction from a grandmother, while a sister effectively brought them up for mothers were busy with other children. Fathers were usually distant figures. Eight had lost one parent and three had lost both: all but one of these told grim tales. Of the fifteen who grew up with both parents, several commented on conflicts between their parents and of infidelity, drunkenness and violence on the part of their fathers. As in Cuernavaca, most mothers were described as patient, enduring and long-suffering. Several authors argue that Mexican mothers bind their children to them and emotionally exclude the father from the family (Lewis 1959; Coberly 1980; LeVine in collaboration with Sunderland Correa 1993): our data would appear to support this interpretation, although some fathers protect their children from their mothers' anger (see p. 161). There is still much variety between individuals and, from our limited interviewing, much remains obscure, and to be explored.

BEING GENDERED

Introduction

What is it to be male or female in these pioneer regions? All the older women are immigrants, from differing cultures and from across most of central and southern Mexico, yet they have very similar views on what a girl, boy, woman or man should be. All should work hard in their sphere. To be a man is to fulfil family responsibilities, to be a good provider, to be serious, responsible, proud, to have authority over his children and to keep his women chaste (wife, daughters, sisters). It is also to be active outside the home, to drink hard with friends and to be promiscuous (but not to embarrass his wife). A woman, on the other hand, should be pure, chaste, submissive, patient, obedient and centred on her children and the private sphere. As in Los Robles, Cuernavaca, the ideal woman is capable of almost infinite emotional and physical endurance (LeVine in collaboration with Sunderland Correa 1993). In practice, the 'ideals' are multiple, conflicting and personalised, but it seems that to be an ideal woman is to develop a capacity for suffering and survival, while for the fulfilment of his role as provider the man is dependent to some degree on the wider economy. A man can be defeated by others, a woman by herself, so that in a sense men are more vulnerable to frustration and insecurity.

These ideals may be far from the real lives of individuals, but they are reference points in all lives and a central concern in writings on gender in Latin America (e.g. Paz 1950; Díaz Guerrero 1955, 1974; Fromm and Maccoby 1970; Nash and Safa 1976; Chant 1984; Pescatello 1976; Cubitt 1988; LeVine in collaboration with Sunderland Correa 1993; Radcliffe and Westwood 1993).

We lack insights from younger women. The women who told us of their lives ranged in age from 33 (Victorina) to 85 (María de Jesús) but we were less successful with young women. Girls before puberty were often very forthcoming, but we found no young woman with the gift of narrative or the willingness to display it to us. Many are very silent and passive with outsiders. The rules are changing, but we saw evidence of this only among the mature.

Why be a woman?

In every 'workshop' (see p. 55) with women, we asked each subgroup, 'Do you like to be a woman? Why?' These workshops were public occasions, which must have been inhibiting, but they were also cheerful social events where some vigorous and unexpected comments were made, often to laughter. We heard what these women were prepared to say in public, with outsiders present, about being a woman. The same themes crop up everywhere.

'I like to be a woman because of . . . ' (89)		'I don't like to be a woman because of . . . ' (22)	
The children	25		6
The kind of work	18		4
The house and home	18		2
Dressing and making-up	12		0
Less work	10	More work	2
Being a mother	6		0
My husband	6		1

In all communities, children are represented at the same time as contributing greatly to the pleasures of being a woman, as a source of suffering, physical and emotional, and as a restriction, because they are women's concern.

We don't like to be women, because we don't go out for pleasure like the men, and the woman has to look after the children and the man doesn't.

(Tacaná)

I don't like to be a woman, because I have no freedom to go out to work because I have to look after the children.

(Independencia)

We don't like to be women because childbirth is hard. We suffer with the children, but when they grow up we want to have another.

(El Tulipán)

Men do not rank highly. We think this may be because it is embarrassing to say in public, 'I like to be a woman to live with my husband and be happy' (said in El Tulipán), and easier in private, in conversation and life histories (Chapter Eleven p. 183).

It is of central importance to have a good home and the garden is also important as identified in all workshops (see also p. 63): 'We like to be women because God made us so and because He gave us the privilege of making homes on this earth' (Tacaná). Specific pleasures of 'femininity' are clearly identified:

We like to be women because we know how to dress, make up well and do our nails.

(Tacaná, the very poor community, carrying all water over a kilometre)

We like to be women because we can dress the little girls with girls' dresses, and with different dresses.

(El Tulipán, the richest community)

Work is described favourably: 'Work is a blessing which God gives us as women: we wash, iron, mill, weed, sow, go shopping, go to the fields and care for the children. Because all the gains are for the children' (Tacaná). Almost everyone speaks well of the work in and around the house, but opinion is divided on farm work. Disagreement arises as to whether women do more work than men, or less. Most think men work more, but some think that women's work is boring, or invisible, or unending. Five comments presented women as superior, as having more wisdom, more creativity, more ability or better taste. Women identify very definite life satisfactions, always sought through those for whom they care. Yet, as we have seen, they are not unaware of negative aspects of women's lives – the suffering, the danger, the restrictions. This is true even in the remotest places, but it was in Independencia, which has the most television and has also had a woman 'facilitator' working with women, that the greatest number of negative points were made about being a woman. We think that consciousness is being changed here by greater contact with Mexican urban values, particularly through television. Overall, 'We all like to work in the house with the children and to make ourselves pretty, and to think more than the men' (workshop, El Arroyo).

Advice to girls

In the course of recording life histories, we asked the women what advice they gave to their sons and daughters, and what advice they would give to women in their own position.

95

What advice do you give your daughters, Doña Aurora (53, Plan de Ayala)?

Well, that on that day that's coming when they marry, they should be good wives, isn't that right, who respect their husbands, and behave well. That they must never learn to take what doesn't belong to them, only what they have worked for. That we must eat and live from our own work, and that they must work honorably, so that no one can find fault with them. That's how they can live well, because, if you behave well with everyone, God is like a friend.

Consistently our subjects express the ideals of respect, good behaviour, respect for property, hard work. Along with these go obedience and courtesy, but there are also warnings of the risk of deception and pregnancy (see Chapter Nine, p. 157). One assumption underlying many of the life stories is that being alone with a boy is being expected to sleep with him. Another is the expectation of marital breakdown. Of twenty-six women, only fifteen grew up with both parents. Duty is important. In Susana's words (68, Sor Juana):

Well, I tell her to be obedient with her husband. On the day he's in a very bad temper, 'Pay no attention to him, keep your mouth shut, never quarrel with him.' Because they have four children and I don't want those children to suffer through her, through her not wanting to put up with her husband. She must put up with him. Because I wouldn't want to have a daughter who left home, who quarrelled with her husband and left him. And we have women who can't stand being alone. Then they think of finding another. But although they find a husband, the children don't find a father. Isn't that right? Impossible for the children. The stepfather is an enemy to the children who are not his own. I – we – suffered poverty to bring them up, but I never put another father over them.

Stepfathers are typecast as evil. (Of our six respondents with a step-parent, five were maltreated, and one raped by her stepfather.) Victorina's (33, Independencia) more urban, international view is extraordinary, even for a young mother in this society:

There are lots of things they set out very well in the books now. So I say, 'Look, as to sex, well, I'm never going to stop you having a boyfriend or anything, but it's one thing to have a boyfriend and another to go on to sex. Sex is beautiful, as I tell you, it's beautiful when you enjoy it, but it has its risks too [. . .] You could get pregnant, that's one [. . .]

'It doesn't make sense for a moment's happiness or satisfaction to cut short your life, cut short your career', I say. 'You have the idea of studying and for a moment of pleasure then in a while you can't, for the responsibility of the child which comes whether you want it or not. It didn't ask to be born and it comes into the world, so it has to be born and you would be responsible.

'No, you go on studying, you can have your boyfriend and everything, and friends, but keep your distance. Then, later, when you have a career, when you are trained, if you want to marry, no one will stop you and you will have the satisfaction of being trained. So go on, and help your husband the best you can.'

Yet Victorina has not spoken to her teenage sons about sex! Other women with daughters still in school have a similar concern that they obtain qualifications first and marry later. In choosing a partner, a girl should choose a good person who likes to work hard, and be in love – but not too much, or he will not respect her. 'Respect' is very important. Girls must always take care of themselves, not let themselves be fooled. They are vulnerable. Antonia (47, Independencia) wanted another boy, 'because girls suffer a lot.'

Advice to boys

Again, Aurora (53, Plan de Ayala) sets the tone:

The advice I give to my boys is that they should behave well, that they should respect from the least to the greatest, so that later they as people, as good men, should never fight with anyone, so that they should have good friends [. . .] that they should work, that they should teach themselves to work, that they should teach themselves to do things so that they won't suffer later. So that later they will prosper. If they don't teach themselves to work, there is no success. As the proverb says, 'to work is to thrive'. Well, that's the advice I give them.

Working and not fighting are constant themes, and so are never taking that which belongs to another, avoiding drink and cannabis and, as for girls, respect. Studying again ranks highly – women fear that their sons will know poverty as adults – and friends are important. Boys have to be told how to treat their wives, as by Julia (44, El Tulipán):

I advise him that he must love his wife and not treat her badly. A man seeks his woman not as a slave, but as a companion. 'This is not the old days', I tell him, 'when the man had the authority – no, the woman must have her authority too because now the woman is also free to decide as well – it isn't the man who is going to decide everything. The woman must decide in married life as well as the man – what are we going to do, how many children are we going to have' [. . .] He must have this with his wife, he must have friendship.

Boys must also be protected. Imelda (55, Naranjales) thought of moving to the town when her husband died, but feared that the move would endanger her sons' morals; they are safer working in the village.

Advice to women

Women must work, work, co-operate and not gossip. They must help their husbands, and must keep them in order to secure an income and a father for their children, however unsatisfactory the husband is (see Chapter Eleven, p. 186). Women are vulnerable. Adolfina (44, El Arroyo), with three sons and a daughter, intends to leave her *ejido* plot to her daughter (who is still single and studying), 'Because she is a woman [. . .] because if her husband leaves her, she'll have some resources.'

The fear of desertion is also a reason for having a skill or profession, as identified by Guillermina (66, La Corregidora):

Well, agriculture is difficult for a woman. I never, never liked agriculture, the sun makes me ill, it's not for me [. . .] I think a woman today should prepare herself, should study to be able to get ahead and not dedicate herself to her house or to agriculture. Because a woman who doesn't know how to read, what can she do? She can do laundry, she can do other people's ironing, she can work on the land and so on. When will she get ahead? When is she going to build a house making tortillas or taking in washing? Never, because all they can earn is a pittance and they can't do anything with it.

Lilia (41, La Corregidora), who had never finished primary school, had this very experience, resorting, when she was deserted, to laundry, cleaning and leaving her children with her sister. Her advice would be:

I'd tell her: 'Life is hard and, now, yes, as they say – tie your skirt up safely and get to work, because you don't do everything in a minute. Get to work, whether it's washing, ironing or selling clothes, looking after children in someone's home, selling tin plates – because there are people who come here and give you plates to sell. And, all alone, you open a space.' I'd also say, 'Don't get sad, keep your spirits up, pray for God's help. Always work cheerfully, by way of doing something, because if you get sad, you don't get anything done and you can get the vice of drunkenness – we have to arm ourselves with courage now.' Pastoral work helped me a lot, being an activist.

But Lilia had no time left for her children, and they ran away.

Guillermina (66, La Corregidora), the entrepreneur, has strong views:

I'm going to say to you that if there are small children, you have to put up with it for them, to let them grow so that they can understand why you are going to leave your husband, because there are children whose parents separated when they were small who blame their mother when they grow up. And worse if there is a stepfather [. . .]

My children themselves told me that I should leave him, because there was nothing left to do, because they saw how he hit me . . .

But as I say to him 'from the street to here is your life, but from the door inwards you must behave with respect. Before, I was a different woman. Now it's all changed. Before, you hit me for anything, once you dragged me from one corner to the other by the hair, only because you get angry, for no reason'. But now, the last time, I got angry too and I said 'Now you are bored with hitting me, now I'm going', and I got a knife out and said 'Don't you dare come in because I shan't hit you, I'll just give you one with this and I'll finish with you', I told him [. . .] The thing is, if you let them hit you the first time, then they'll take it as a joke [. . .] so you mustn't let them. And I was slow, but now I don't let him.

Women alone

According to Juliana (75, Plan de Ayala), a woman alone suffers much:

When my dear mother left my father's control, how she suffered. She couldn't bear it [. . .] She went out to work with my little sister, who was 2. She went to Los Tuxtlas to work, and told us that she had a terrible time, because, she said, they used to give her two little tortillas for her and the child to eat. She gave both to the child and went hungry. When she used to talk about her life [she weeps] – she suffered dreadfully.

Celia (60, La Planada) told us how difficult it is to be a widow:

And it will be more difficult soon now, very soon. In the first place because my sight is failing and second because I think every day about how I'm going to manage alone. It's true that there are those two boys here, but they are my grandchildren, not my children. What am I going to do? Tomorrow or the day after, they will be grown up. I can't know what direction they will take, whether they will go, also whether they will look for a partner, whether they will go away. They are going to leave me alone here.

Antonia (47, Independencia, born in Mexico City) feels a need for a man:

Well, there are times that, yes, because you feel alone and then – you need embraces, don't you, and so on, but – [Antonia's mother interjects:]
'Why do you need an onion to cry? No, I'm telling you that the couple is essential for this – for companionship and also because the body needs it, more than anything.'

Antonia has fought alone. She regrets having been unable to breast-feed her children, being a single parent out at work all day in Mexico City. Now, she has inherited an *ejido* plot from her father, but,

Since my father died, I don't go even to the meetings, only the required ones, because they are all men [. . .] here, anyway, if you say anything as a woman, they call you crazy. They don't pay any attention to us, least of all when it's a men's meeting.

However, Antonia is unusually positive about her situation. She has been held back while the children were small, but she has seen herself through adult secondary education and will get ahead now, for she is free to decide, while many women around are discouraged by their husbands.

Leisure

Women narrate their memories of work as children with enthusiasm, but say little about their work now. They just do it, from 4 or 5 a.m. to 10 p.m. and sometimes more. Some say that younger women now work less and a few women now watch soap operas on television, but from the youngest at 33 to the oldest at 85 the leisure of most is the same:

And you with so many children, Doña María [La Corregidora], how did you amuse yourself, how did you rest?

Well, look, I worked a lot. Believe me, washing, ironing for all of them, I worked hard and so did my daughter. But eventually daughters-in-law came, then I rested a bit, yes, I rested a bit.

What do you like to do for fun?

Me? Well, anything, with the children. I love children, yes, I love children and I especially love them to play, not to be sad. It pleases me when children are happy – yes, my grandchildren too.

During the workshop discussions, there were many comments about how 'the children are our rest'.

Women earning an income

On the land

Women's limited opportunities to contribute to production or earn an independent income are an important part of being gendered in pioneer areas. Most women have never worked in the fields and, of those who do, most give it up as more children are born (Table 2, p. 63, Figure 1, p. 64). Young children and youths make demands on time which would astound many African women who work long days in the fields. Susana (68, Sor Juana) picked coffee for a wage when her first children were young, but then, as a pioneer:

Doña Susana, what did you do when you came here?

The cooking. Making their food and washing. And there were so many of them, I had too much to do, because there were so many [. . .] They went to work, they did. And I was skilled in the kitchen, waiting for them to eat, making their food and getting their clothes ready, because you know, in one day, boys – well, they want their clothes now, now! They get in from work and they want to go out and they don't want to go out as they come in!

In the poorest community, Tacaná, and in indigenous families, more women 'help' in the fields (see Chapter Eleven, p. 185). Women or children must take men's food to the fields and women speak of having done a little work 'to help'. In poor families, they also carry the crops back. Some women, usually widows, are *ejidatarias*, the rights in a parcel of communal land being vested in them (Chapter Eight, p. 144).

Rosario (34, Plan de Ayala) is one of the exceptions who take great pride in agricultural work and simply take their children along. She despises women who do not for she worked in the fields until a fortnight before her last confinement. When the baby is grown a little, she will find a girl to look after the children and go back to the fields. She takes pride, too, in her livestock – fifty chickens. She tells the children – and us – repeatedly: 'You have to work and have your own [. . .] I know how to clear land with a spade or a machete. I know how to hold the axe to fell a tree [. . .] What I never liked was carrying wood.' The labourers whom the couple hire to work in the chilli fields obey her better than her husband, she says, because she can do the work faster and better than any of them.

Off-farm work

Nationally, as wage levels have fallen in real terms over the last ten years, women have struggled to hold up family incomes by joining more and more in earning (Arizpe *et al.* 1989). Pioneer women are eager to enter paid work, but their wide range of traditional activities generate little income (Chapter Nine, p. 159, Chapter Ten, p. 171). The main paid skill for women is midwifery, mostly untrained (Chapter Eleven, p. 188). All rear chickens and perhaps pigs, which are seen as walking savings accounts. Some sell cosmetics or plastic utensils door-to-door. From Independencia, where women have been trained to embroider, women go to Escarcega to sell napkins, pillowcases and tablecloths, although the returns are very low.

These minor activities rarely yield enough money to live on, so a woman alone without land will have to leave any small community. When Imelda's husband left her (55, Naranjales),

> I went on working, working, travelling to seek shelter anywhere, travelling to beg. Sometimes the people where I begged shelter wanted me [to work], sometimes no. Then I'd have to go somewhere else to beg for a place, just to manage to bring up my children.

A disproportionate number of our life stories come from women who are supporting their families. We think that their confidence in telling their life stories to outsiders relates to their role in public space. Five widows and three

101

single parents support their families and two others earn more than half the family income. Four of the old are supported by their children, while the remaining twelve depend primarily on their husbands' income.

Conditions in paid work are usually poor. Raising a family by cleaning floors and working at casual jobs is desperately hard, as Lilia and Irma (41 and 34) have found even in the small town of La Corregidora. In prosperous El Tulipán, Yolanda (44), abandoned by her husband, cooked in a nearby rice mill from 6 a.m. to 11 p.m. When she left that job, she made soft drinks, tortillas and dumplings for pay, getting up at 3 a.m. to start selling at 7 a.m. Now, a man from the town advances chicks to her to raise and sell but, as chickens are traditionally given to one's children or sold on credit, it is hard for her to conduct a rigorous commercial enterprise. She is considering renting an electronic game and buying a small refrigerator to sell bottled soft drinks.

Entrepreneurs

Carmela (p. 148) and Antonia (47, Independencia) have shops as well as their *ejido* plots; Celia (45, La Planada) has a chicken farm and all the women have to be entrepreneurs with their home gardens. Guillermina (66) has been the most successful, trading fish to Mexico City. She came newly married to La Corregidora when it was only a few shacks:

> Well, he was really poor. It was truly from one world to another, because my father [a teacher] had property but I didn't know his place nor where it was nor what it was like. And then, after I married, well, he brought me here. Then I lived a very miserable life in truth [. . .] I set myself to work at his side, yes, in trade. I started with him and said to him, 'Oh, look, you're letting someone else earn the money I need for my children. Why don't you sell the fish [you catch] to me?' [. . .] And I began to buy fish and to sell it [. . .] And so I kept my children. And I used to buy *barbasco* [the yam used in the manufacture of contraceptives] [. . .] I travelled as you [the interviewers] travel, but without losing the respect of my husband, because, well, you know the value which one has as a woman? Truly, you make your own value, there's no need for anyone to protect you. You travel but you know that it is as if your husband is with you, his presence is behind you [. . .] He didn't like my travelling. He was macho, one of the men who want to feel they have total control [. . .] And I said, 'No. All right, you behave as you like, I'm going to get ahead. I don't like this misery. It means there are days when there's nothing to eat. I can put up with it but the children can't' [. . .] I began to buy fish very young, at about 23 and without any experience [. . .]
>
> I never had a servant [. . .] I got up early to have lots of animals [. . .] From what I made, we all ate, and what he made, he drank – but I put it to him, 'Choose either your children and me, or your vice. Certainly, I can

work and I can keep my children.' And he gave up drinking. . . I can tell you with pride that it was I who built this house, because he brought me to a miserable wooden hut, and I was able to get ahead and make my own home.

However, despite their varying degrees of success, women entrepreneurs and organisers are always conscious of being away from the children and being in public space.

Public roles

Three of the women who told us their stories were particularly active in the public domain. Tomasa's (42) husband permits her to travel, as the local President of Rural Women in La Planada, and is supportive:

They asked me, 'Where is the baby?' I burst out laughing, and said, 'Look this will make you laugh, but go and see. I brought my little girl to the meeting and my husband stayed with the baby [. . .] he's staying alone with the baby until seven o'clock when I'll get back.' If I'm going to go out, I know what responsibilities I have.

Isabel (51, a widow) can neither read nor write but runs a successful maize-grinding business. She is about to be nominated by La Corregidora to represent municipal authority there – a rare achievement for a Mexican women (Robles *et al.* 1993). She has had one woman predecessor, who built the school and the 'telesecondary' school where teaching is assisted by television: 'That's why they often say that a woman has more ideas than a man [. . .] and yes, it's true that a woman has more anxieties about where she lives than a man.'

Adolfina (44), elected leader of the *ejido* of El Arroyo (another very rare achievement for a Mexican woman), also has to travel:

When you, as a woman, go to the offices in Palenque, how do they treat you?

They treat me well. Because, I think, it's because I'm a woman that they respect me, isn't that it? And I get there and greet them politely and go straight to what I want to deal with.

Do you like going to the offices to do business?

I love it! [She laughs.]

Adolfina is an *ejidataria* in a powerful family and her husband is a teacher. She organised a group of relatives to buy 460 head of cattle with bank credit. She plans a kindergarten, a clinic and a playing-field, improvements to the access road and ponds for fish and to water the cattle. She claims to have persuaded the Department of Family Integration, through the Governor's wife, to build the big training centre in the village (see p. 71). Adolfina has deployed her land, family and skills to build power in the village and manipulate the bureaucracy to remarkable effect.

PARTNERS AND MARRIAGE

In these communities, almost all women form heterosexual relationships, bear children and eventually set up a new home with their partners. A very small number of men live together and are referred to as homosexuals, but no women. We gained no insights into the experiences of lesbian women in rural Mexico. There is a big range of heterosexual relationships, despite a ruling ideology of female purity and male power. Traditionally, women married and were faithful and a few widows remarried. Most couples do have a civil marriage, a church marriage or both, but a few never marry and a minority change partners. Some women described at length the ways their first relationship had developed, while others dismissed it in a few words. The many tales of marriage ceremonies always include second thoughts, doubts, distress, expressed around the loss of family and friends, the transition to a new life.

Not all see a binding marriage as a safe future. Adolfina (44, El Arroyo) is a forceful woman:

> Only a civil marriage, yes. Up to now, nothing more [. . .] My mother was very Catholic and so she wanted a church marriage. But I said to her, 'Who knows? Because possibly', I said, 'I shan't be happy with him, and then you see there would be problems.' And in those days, they said that anyone who married in church had to put up with whatever happened. So I didn't.

Age

Many of the women who told us their stories married very young. Imelda (55, Naranjales) married at 13, 'because we didn't have a mother'. Nationally, the mean age of marriage for girls is now over 20, higher in urban areas and lower in rural. Here, in the more prosperous communities in the pioneer areas in which we worked, more open to urban influences, the age of marriage has risen, but otherwise early elopements are still common. It is not necessarily a recipe for disaster, for Isabel (51, La Corregidora) is still cheerful about her elopement (she pretended to be doing the laundry at the stream and slipped away). Irma's story (34, La Corregidora) is very different as, despite formal marriage, she is now a single parent:

> For me, it was the only easy way out I could see, I thought that if I got away I'd be happy, but, what happens? On the contrary – it was worse [. . .] [having been abused by her stepfather] I wasn't a virgin when I went with him, and that's what he quarrelled over with me [. . .] Because I told him the truth from the beginning, but, over and over, when he was drunk he remembered and reproached me for it [. . .] At first it was good, at first it was fine [laughter], everything was love, but then later, then my sufferings began.

Today, girls of 15 (unlike those in Cuernavaca (LeVine in collaboration with Sunderland Correa 1993)) have still almost all left school and are still usually secluded and chaperoned until they elope or marry.

Sex

Most women were willing to speak to us about sexual relations in response to a leading question, some very favourably, but usually the favourable comments are brief. Ignorance is cited as a barrier, as by Victorina (33, Independencia):

Listen, Doña Victorina, a very intimate question, between women. How did your husband treat you, sexually? We all want and desire and also enjoy it, don't we? Did you enjoy it, did you like the way he treated you?

Look, before, no, because, to put it straight, he used almost to take me by force – that got me down. It's more [. . .] now I'm enjoying more what I didn't like when I was just married because before that, well, I remember when it was our first night, no? He almost took me by force and blows [. . .] as I wasn't ready for it, right, well, I didn't know how to respond so it didn't seem to him – and we went on like that, with him having no skills, not knowing how to respond, how to make me feel [. . .] Now that's changed me a lot.

For me, sex is something natural, very beautiful, and should be done with respect, but at the right time, when my body calls for it.

(Lilia, 41, La Corregidora)

Lilia and others report a loss of interest with age, which seems to be the respectable thing to say. Guillermina (66, La Corregidora) illustrates some of the impact of the external world as she relates her own experience to her recent reading:

Well, yes, he loved me but he was very coarse. Because I read, now I know what sexual relations really are, and what value one has as a woman, because reading you find out who you are. And, reading, you come to think, who is he? And then at times you don't find in your husband what you want in a man, isn't that so? [. . .] A man should know what his wife is like, hmmm! Study her and make love as she likes it, to leave her satisfied, because that way – you, what could you seek from someone else? Think about it. And that's the cause of marriages breaking down, in many cases, that the woman doesn't get ready, doesn't read a book, and the man much less, he just feels the sensation of being a man [. . .] They make a woman work hard, and then want to use her in some clumsy way [. . .] Since I read those books, something really big happened to me – I woke up.

Most women look forward to the menopause, but many report real ill-health at that time with heavy bleeding, big physical upheavals, forgetfulness, hot flushes, irritability, nausea, headaches. Guillermina (66, La Corregidora) has hormone treatment, Antonia (47, Independencia) has pills, vitamins and injections and Susana (68, Sor Juana) lay in bed for months with depression.

105

Childbirth

Pregnancy is usually seen as a joy (Chapter Ten, p. 170), but most births take place in a context of hardship. There may be no help, or more often an untrained midwife.

> OK, that one, my eldest daughter, was born. The next day I had to get up, make dinner. I didn't have anyone to do anything for me, nothing, nothing in the house, not even him [her husband] to get me a bucket of water or anything [. . .] we were so poor that when I got up the next day, to cook, there was nothing, nothing.
>
> (Yolanda, 44, El Tulipán)

(The World Fertility Survey reported that in Mexico sons were preferred over daughters, but this was not clear in our findings.)

Fertility

Contraception has been a government concern in Mexico for less than twenty years, but it is a live social issue. In the rainforests, fertility, or completed family size, is very closely associated with work opportunities for boys (Arizpe *et al.* 1993; Townsend with Bain 1993): fertility has declined dramatically over time, especially where most land is in pasture. For most people this decline in fertility is stressful. Even in the general survey, many women talked to us about contraception, although there is an ambivalence, for many give medical reasons for sterilisation, as if medical grounds are more respectable. Most women who talked to us are Catholics and the Church in Mexico is very hostile to birth control, but women rarely mentioned this. In other, very different communities in Mexico, people are investing more in children and having fewer of them (Rothstein 1986; Simonelli 1986). Although Victorina (33, Independencia) refused, her husband wanted her to abort their third child with an illegal injection 'because he hadn't got the means to keep us'.

Sterilization by tubal ligation is the preferred contraceptive method in Mexico and is available to women who are over the age of 25 or who already have three children (LeVine in collaboration with Sunderland Correa 1993). It is widely used where we worked and the hospital in Palenque offers the operation free.

> I was sterilised because we couldn't cope any more with so many children. What could we do? Fill up with children – and the day they had to go to school, not be able to send them? So I had the operation [. . .] I decided, and then, when the chance came again, a big crowd went to be sterilised, because they saw that I hadn't died.
>
> (Josefa, 42, eight children, La Planada)

Injections are also popular and some use the pill. Vasectomies are not considered: 'It's machismo: he has to go out and show off' (Ernestina, 38, Independencia). A girl known to take the pill is disgraced. In some cases, the man decided on sterilisation and the woman accepted, against her own wishes (Chapter Eight, p. 147, Chapter Ten, p. 174). The sterilisation of Yolanda (44, El Tulipán) led to bitter resentment. When she had had three children, the rancher who employed her and her husband took her to hospital to be sterilised 'because he did not want to see me suffering, or the children'. Her husband was extremely angry, but she feels 'we would have gone on burying children'. Her husband then resented having to wait six months to resume intercourse and took up with another woman: 'Now I was, I was no use as a woman any more.'

In other cases, women were sterilised without their consent and against their will, like Lilia (41, three children, La Corregidora):

> They sterilised me after I had the twins! When I was on the table, the doctors asked my husband, 'Shall we sterilise your wife?' He said 'Yes'. The doctor told me when he took the stitches out. Well, I wanted to kick that doctor, because I wanted to have more children. I'd always had the idea that I was going to have twelve children! [. . .] Because my mother had so many, I wanted to have many children too, because I think it's beautiful to have lots of children.

When Lilia's husband deserted her, her sterility reduced the chance of another partner. Conversely, Leila, in Balzapote, insisted on seven children although her husband wanted to stop at four.

We heard a great deal about side-effects of sterilisation, ranging from headaches to haemorrhages, which are a very live issue. Two told us in their life histories of a long chain of troubles before sterilisation. Josefa (42, La Planada) reported weight loss with contraceptive pills, sickness, diarrhoea and fever with injections, fluid retention with a coil. (Several complained of unwanted pregnancies despite contraception.) Finally Josefa became pregnant and was sterilised after the birth, 'because if not, I wouldn't now be seeing her grow up'. Ernestina (38, Independencia) has had twelve pregnancies. Her husband wants her to be sterilised, but she is afraid, for her sister's brain, she says, was 'affected by the anaesthetic', and her sister-in-law 'was left with a really nasty pain'.

There are still the women who cannot get the contraception they want because their husbands refuse to allow it. Lucía (39, Jasso) was warned at her last delivery against another pregnancy:

> I thought to myself: 'I'll never be with him again.' But, as he drinks, he forced me, and that's why I'm like this now [pregnant]. Because he came into my room, and I was resisting him and my daughters were telling me not to let go, so I locked myself in the room but he was angry, and the girls were frightened and I don't like it when the girls realise what's happening. Only the eldest, I can confide in her, she knows what goes on [. . . she is

married]. So this is why, you see, I am like this, not because I want to [. . .] If I could have chosen, I would have stopped with my third or fourth child [. . .] I tried to get them [the injections] through a friend of mine, because she too suffers because of her husband who also beats her and rapes her [. . .] But at least she's got money [. . .] Yes, because she works. She sells some fruit, lemons, she's got a lot of lemons and something else – I can't remember what, but she sells a lot. Oh, yes, coconuts. She sells coconuts. So she's got an income and she can afford the injections [. . .]

There is a clinic where you can go to discuss these things, but you need your husband's signature [. . .] Yes, the husband must sign, or else they won't see you.

Despite the pressure for smaller families, children are still valued (Chapter Eleven, p. 191). Orphaned children are adopted, as by Aurora (53, Plan de Ayala), who claims that the baby was given to her by its mother in the street.

Infidelity

The way the women talked to us portrays their purity for they speak of their partners' jealousy and accusation, but never of their own infidelity. Many spoke of their partners sleeping around, particularly in La Villa. Victorina (33, Independencia) is, unwillingly, considering a divorce. Her husband (the elected leader of the *ejido*) has been neglecting her to be with a girl who has borne him a child, but with whom he does not want to live. Victorina has tried to help the girl with money. Victorina is also severely beaten:

And after all that he has done to me, I love him very much, I tell you. I think that has also helped me put up with it, because, I tell you, apart from the love of my children or the love for a father for my children, I love him very much.

Nevertheless, we heard far less about the 'other woman' than did Sarah LeVine and Clara Sunderland Correa in Cuernavaca (1993). It may be that if we had been able to undertake repeated interviews with our respondents, we would also have heard more of her. In Mexico City, Irma Saucedo (1993) found that most women only went to a women's refuge for help when the husband took another woman – that 'they accept everything but betrayal'; drunkenness, violence, rape are endured, but not infidelity. Here, violence is the much more central concern but there is still often a difference in women's tone between distress about domestic violence and anger at betrayal.

Prostitution, with a very small number of women earning a living from casual sex, appeared to be present in most communities, but our short visits gave us no insight into the relations around it. No woman ever told us that she had earned from prostitution, and none recounted any casual sex. All spoke only of long-term relationships. They do speak of other women who live from prostitution and of others who are regarded as a threat because they are known to be even a little promiscuous.

Rape

It surprised us that all the stories of rape of adult women were of conjugal rape, not stranger rape, just as all child abuse was by stepfathers. Drunkenness is normally implicated. Complaints were rarely articulate, but everyone is aware of the issue, even if they do not suffer. Ernestina (38) moved to Independencia because she was molested by her father-in-law:

> And once I grabbed a log out of the fire, a lighted log, and had to hit him! Because he made me so angry. And I told my husband. I think he was afraid of him, because he didn't say anything to him.

Countries differ greatly in their legal attitude to marital rape: in India, marital rape is in no circumstances a crime, while in Sweden and, now, in the United Kingdom it is a crime even when the couple are living together.

Drunkenness

Overindulgence in alcohol by the man (called 'alcoholism') is women's most commonly reported source of misery, and a long-standing one. Chavela (85, Sor Juana) lost her parents and married young:

> I got married at 15. And God had pity on me, that I didn't have more than one child, because my husband turned out dissolute. He drank a great deal [. . .] I ate because I ate where I was working. And he also drank everything he made from the horses, and there you have it. So my life was very miserable, but later I left him [not for another man].

Men's drinking is a popular subject.

> I want to feel good. I want him to change, for his sake too, and for the children's. But he says he'll stop drinking when he feels like it, not when I tell him to. Before, he couldn't drink raw alcohol [sold illegally]. Now he does [. . .] I'd like to get some advice, know how to behave with him. How to treat him. Maybe it's because I don't know how to behave that he feels more and more inclined to drink [. . .] If he at least smelt clean – but he smells horrible and I'm revolted by it. And he says, 'I drink because you don't love me' [. . .] I don't like feeling like this. After all, he's the father of my children [. . .] One of my daughters says 'I won't feed that drunk.' And I say, 'Child, he may be a drunk, but he is your father' [. . .] When I go to the river to wash, or when I'm with the baby, the children go and feed him. But they don't like doing it.
>
> (Lucia, 39, Jasso)

Drunkenness is often implicated in marital rape. Sobriety is a part of women's conscious virtue, as Aurora boasts (53, Plan de Ayala):

It's me my children trust, not their father, because their father sometimes drinks and whenever he's had a few, it's not the same as being in his five senses. That's why it's me they trust [. . .] Sometimes, the men squander everything. It's up to me to take care [. . .]

Doña Aurora, if you were to be a widow, what would you do?

Well, I'd try to bring up the children. I've already done it, because my husband likes to drink, and when he drinks, and when he gets hold of it – he gets hold of it! Up to six days [. . .] He doesn't bother with his family, whether they eat or don't eat, he comes in and eats, and knows nothing about the rest.

Drunkenness is indeed an economic problem for women:

Most men get their harvest and say 'Let's go here', and spend it, get drunk and are back where they were [. . .] They leave their families without food, that's why there are some little underfed girls going across there [she points].

(Antonia, 47, Independencia).

Some religious sects ban alcohol, which is very welcome to women such as Carmela (p. 152).

Gloria (47, El Tulipán) is unhappy because her husband is paying attention to the woman leader of a non-Catholic sect, but she is delighted that the sect has required him to give up drink. They suffered on arrival in the village:

Everything went well, and then he was working for the local government and doing well. Only they sacked him for his vice, because he used to drink [and drive . . .] I told him, 'For this vice of yours, we're all going to suffer, because, the children too, the day will come when we have to eat and there is nothing [. . .] Because they have to eat', I said, 'even if it's little snails, even if it's what they call, to trick them, little tortillas with lard and salt, but they have to eat something.' And so it was, with three little tortillas and their bottle of pop and then they'd say, 'Now I'm full, Mummy', and I'd say 'Thank God, now you are full, now we'll wait for the afternoon and see what God gives us to do.' And so, you see, he went looking for work, also my son, for work which would help us, because when he works – my poor children – he gives it to us, but, oh! *señorita*, there isn't any work.

The tales are variations on a theme:

[The men] get home drunk, ready to beat their wives. And not only that, they spend the housekeeping money on it. We women put up with it somehow, but the children, their poor children – what crime have they committed which could leave the family without anything? And it's the woman who suffers, because there's nothing with which to feed them, and that's the worst of it [. . .] Oh, God, this is why I give thanks that my husband doesn't drink.

(Josefa 42, La Planada)

Eric Fromm and Michael Jaccoby (1970), working in a highland town, concluded that the impulse to drink arose chiefly from the frustration of men's sense of manhood. To refuse alcoholic drink is seen as a failure of manhood (Natera 1983). Change has compounded the problem of alcohol abuse: in the late 1970s the national sale of alcohol in rural areas rose by 50 per cent and it became one of the principal health problems (Natera 1987). There are five bars for every school in Mexico and, in the population aged 10 years and over, one in 100 is an alcoholic, while in new communities (such as those we studied) criminality parallels the number of bars per 100 inhabitants (Velasco Muñoz 1983). The National Health Survey found that 14 per cent of Mexican men drink to excess (Solache 1990).

Most women drink little or never, and never get drunk, but Irma (34, La Corregidora) did report getting into a drink problem when her son was taken by his father:

I drank a lot [. . .] Yes! I worked, and devoted myself to my children, but when the moment came when I remembered my son, I recalled so much about my son, I gave myself up to drink, I drank every day, every day. But then when I used to get home, to the room where I had my children, I'd see my [other] son, who would say, 'No, mummy, don't go on like that, look how we are here alone' – he was tiny – 'now don't go on drinking, think about us, because your drinking won't bring back my little brother' [. . .] Little by little I made the effort and thanks to them began to give up drinking. I thought, I have to get ahead, I have to give my children a better life, and then I began to find a way to do things, to give them better things, clothe them, shoe them and so on until I began to give up drinking. I was like that almost a year.

In Cuernavaca, Soledad (LeVine in collaboration with Sunderland Correa 1993) also drank in the face of crisis, and stopped because the children begged her.

Control over money

Control over money, like control over fertility, is a constant source of conflict between partners. The household budget is managed in a variety of ways. Josefa (42, La Planada) represents the most widespread: 'He works and gives me my week [housekeeping money] to eke out so that it will last us the week.' Some men hand over the 'week' faithfully, but effectively the money is the man's to withhold and a very common cause of conflict and violence. In more extreme cases, the husband handles all the money and the wife has no allowance at all, leaving her extremly vulnerable:

> They [men] don't just go to drink, but also to houses [brothels]. They also spend money that way. Yes, they spend a lot of money [. . .] I don't have anything, not a cent. That's the way he wants it. He has been told that his way isn't the right way, but he insists women shouldn't have money, and who is going to convince him otherwise?
>
> (Lucía, 39, Jasso; see also p. 107)

Juliana (69, Plan de Ayala) was able to sell the orange crop from the garden to pay for medicines which her husband refused to buy, but normally a woman's income is for the family, not for the woman:

> Well, when I sell an animal, we spend it on the needs of the house, on food, because everything is so expensive at the moment that I want to sell a chicken now too. I have one hen which makes me run about a lot, so I want to sell it because I want to pay the electricity – so expensive!
>
> (Petra, 60, La Villa)

Ernestina (38), perhaps influenced by the women's worker who was in Independencia, advises her sister-in-law:

> 'Now when you want to sell an animal out of your yard, like a turkey or a chicken, you have to ask his permission. But you own the turkeys, you raised them! When you want to buy yourself a dress or something, sell them! Don't let him know!' I say, 'it's you who has all the hassle, not him! He has the right to his things from the fields, to his things, he goes and sells an animal, kills deer, and sells it. The money he – drinks'.

Guillermina (66, La Corregidora) gives pocket money to her husband while Adolfina (44, El Arroyo, see p. 103) and her teacher husband, unusually, run separate budgets.

Violence

Fear plays a role in the need for a partner, in the maintenance of patriarchy and hegemonic heterosexuality (compare Valentine 1989, for Britain). Lilia (41, La Corregidora) had been abandoned by the man she had married in church, but after her children left her she became doubtful about living alone: 'I was afraid. People aren't bad here, but, after dark, when the neighbours weren't there, I worried about being a woman alone in the house.' Lucia (39, Jasso) has similar worries about leaving her drunken, violent, rapist husband:

> At least here he's some kind of protection. Because if I'm left alone, a woman alone, the young men will have no respect for the girls. That's why,

112

in a sense, it's not so bad that he's here. They see I'm not alone. I have thought, 'I'll leave with my children. Someone must be able to give me more than he does.' But I'm afraid. With so many kids . . .

But fear also works the other way, for the fear of violence keeps Carmela (45, Plan de Ayala), a widow, from remarrying (see p. 149) Like poor women in the cities of Querétaro and Guadalajara (Chant 1984; Gonzalez de la Rocha 1986), many pioneer women are frequently the victims of domestic violence. They are beaten whenever drink allows their husbands to explode with anger and frustration, yet are usually kept from leaving by pressure from other women and by the lack of an alternative income. Tolerance of infidelity and violence is exchanged for an income and a father for the children:

> We suffered a lot in our first years. On the one hand there was the poverty, on the other, we didn't understand each other. He used to knock me about a lot. [She cries bitterly and at length.] When I was carrying my second boy, he treated me really badly, with kicks and blows. So he was born with a lump here [on the brow]. I remember that when I was very big, ready to fall into bed, I went for the water with two buckets [. . .] I'd taken his lunch early [to the field where he was working] and stayed a bit to help him. We got back at about 1. Instead of sitting down to rest or anything, I set myself to find wood, and after I'd got the wood, I went for the water. I brought the water in my two buckets, to save time, I brought my two buckets of water. Then he came, and sometimes he doesn't give me time to mill [the maize for tortillas] and to make a quick meal and sometimes the meal isn't, the meal isn't ready. 'I go to help you and with the work to get the wood and water and everything I don't, I don't have time –' 'No!' – heaven knows what – 'you are good for nothing!' And then come the blows [. . .] And that time, he got me by the hair and dragged me around and then he gave me a kick. And I was really big [with pregnancy].
>
> (Victorina, 33, Independencia)

Lucia (39, Jasso) had eloped with her husband:

> A year after we got married, Chilo was born [. . .] Two weeks after she was born, he hit me and sent me to bed and tried to strangle me, because the little girl was fair so he insisted she wasn't his [. . .] So my father took me home and said, 'You won't go back to him.' But he [her husband] persisted so much that in the end I went back [. . .] If I stood all that, it's because I loved him [. . .] He was very violent with Chilo. He struck her often. In the end, I had to take her away because he hit her with a machete.

Attempted strangulation is a recurrent theme.

In societies worldwide, it seems that there are men who beat their wives, but that wife-beating is rare in some and frequent in others and the severity may range from mild to murderous (Naciones Unidas 1989; E. Brown 1991). In many societies, such as Iran, India and Taiwan, some 'moderate' chastisement of wives is expected, while in others, such as the United States and Britain, there is supposed disapproval of any beating of wives, but despite this wife-beating is apparently widespread and severe (Campbell 1991). Societies like India and Brazil, in which even wife-murder is condoned, are rare (E. Brown 1991). The immediate 'causes' of violence vary: in some societies, male sexual jealousy is important, but in India, for instance, it is irrelevant to the majority of wife-murders. The stimulus to violence may be trivial (Molino Pineiro and Sanchez Medo 1983), but drugs and drunkenness often play an important role, especially after drinking sessions during which men abet each other and reaffirm their role as breadwinners and heads of their families (Naciones Unidas 1989).

For the working classes of Guadalajara, Mercedes Gonzalez de la Rocha (1986) seeks to explain (but not justify) wife-battering, not simply by lack of education, a miserable childhood or incompatible personalities but as a result of the man's impotent position in the world of work. The private sphere is where the man sees the chance of exercising control and reaffirming his power, and violence is one means of achieving this, consciously or unconsciously. Men, say Sarah LeVine and Clara Sunderland Correa (1993), of the working classes in Cuernavaca, feel that they have to protect themselves against a world that constantly belittles them. We came independently to the same conclusions. Sarah LeVine (LeVine in collaboration with Sunderland Correa 1993: 201–2) sees men's difficulty in the world of work as a phenomenon of Mexican cities, where, she argues, the change over the last half century is harder on men than women. Only among his drinking companions can a man be, 'instead of a poor provider and negligent father, once again a man among men' (she quotes from Elliott Liebow, writing of working-class black men (1967)). As the pasture takes over in pioneer areas, they resemble the city in creating stress for uneducated men, for whom there is little work. At the same time, nationally from 1983 to 1988, the fall in real wages per worker in agriculture was some 35 per cent (calculated from Lustig 1992) and, despite non-wage income from their own production, the rural poor suffered a substantial deterioration in living standards (Lustig 1992). National crisis compounded the local agrarian crisis for those providing for a family in pioneer areas.

Some authors describe Mexico as a society in which physical violence against wives is simply the exercise of a right consecrated by tradition (Lailson 1989). More see it as both an instrument of domination and the discharge of frustration (Gonzalez Montez and Iracheta Cenegorta 1967). Wife-beating is then the main release for the frustrations, tensions and conflicts engendered by poverty and oppression. As everywhere in the world, wives are blamed for domestic violence and marital rape (Lailson 1989) and there are myths that they enjoy it, that they provoke it, that they are unfaithful or unloving.

Resistance

Some women have escaped this cycle of violence (Chapter Ten, p. 173). In general, the woman is supposed to solve it for herself. It is her responsibility:

> The man [is ignorant, uneducated] because he doesn't know what he has, and the woman too is ignorant because, well, yes, from the beginning, when she went to marry, she didn't make the conditions for respect so that she could protect herself.
>
> (Antonia's mother, Independencia)

Some never succumb, like Florencia (77, Sor Juana):

> Well, when he came in drunk – he used to drink a lot – when he came in drunk, well, he'd fall down on the ground and stay there. And once he wanted to do something to me, he hit me, and I told him, 'Now, you don't give the orders, to get drunk, to come and lift up your hand, to want to hit me.'
>
> He said, 'Sorry', he said, 'I don't remember, I don't remember.'
>
> He was drunk, well, 'It's the first time you've got drunk and wanted to hit me and shouted at me, but you just forget it, forget the whole idea!' And now, no, but now he doesn't drink [. . .] because he saw how two died, he saw how twelve died, here in Sor Juana.

Aurora (53, Plan de Ayala) is also proud of her resistance:

> *Did he ever hit you?*
>
> Yes, sometimes, yes, but – I never let him, I stood up for myself! He sometimes went to hit me, but I never let him! [. . .] I tell him, 'I'm not your slave!'

Violent resistance is not always reported as being successful. Irma (34, La Corregidora) left one partner

> Because he treated me badly. That is, he hit me and sometimes even left me without food, without money to buy food . . .
>
> *When he hit you, Irma, did you defend yourself?*
>
> Yes, once! He'd beat me really badly [. . .] One day I was angry too, really, and I thought what to do – well, kill him! Because he'd made me sick of his always beating me and beating me and beating me all the time. And one day he was in bed and I had him covered with his own gun, that is, I meant to kill him. I changed my mind because of my children. I thought of them and leaving them alone, and at that moment I came to myself and thought only of my children.

This was the man who later kidnapped, at gunpoint, the boy they had conceived together and whom she never saw again.

Official action over wife-beating seems to be confined to cases of murder. Doña Adolfina (44) claims that, as elected leader of El Arroyo, she sends for men who beat their wives and gives them advice, but we did not hear of this from anyone else in the village, and she did not speak of action beyond advice.

Action against alcohol

Many women want to see action against the abuse of alcohol, or even against the consumption of alcohol itself (p. 152). In Jasso and El Arroyo the sale of alcoholic drinks has been banned at the instigation of the women. Adolfina (44, El Arroyo) involved the Governor's wife:

> Well, look, I asked the health [office] which is there in the Municipal Council [. . .] and I insisted and insisted and so on, and when the Governor's wife came [to El Arroyo] I put it to her [. . .] and yes! See! It worked! There's still one person who I hear goes on the same, but just yesterday I went to the Municipal President to talk about this, so that he'll send for him to advise him to stop it.

Josefa (42) has been getting women in La Planada together to propose the same thing.

Liberation

Guillermina (66, La Corregidora) has built a commercial enterprise, gained control over her violent, drunken husband and read some popular sexology:

> The important thing for me is that, the woman must work, liberate herself. Liberate herself and not be tied to the man, because in the end we don't find in the man what we are looking for – womanisers, drunks, no. Because their life is very different, it's of the world and you get really ground down because you are really weary from the time you marry.

Lucía (39, Jasso), who has failed to control her alcoholic, violent husband (p. 107), does not find liberation so easy:

> [The nun] used to scold me all the time. But I tell you, nobody knows it better than the one who is living through it, the one who is living this whole thing. Anyone can say, 'This woman is very stupid.'
>
> My mother is always telling me off. She tells me, 'You are just being too complacent, and let him do as he wishes' [. . .] There are still people that live like this, very subordinated [. . .] You want to liberate yourself. But people think harshly of you. You just can't.

Evaluations of marriage

Some do speak of felicity, like Elena (p. 158), Josefa (42, La Planada) or Susana (68, Sor Juana). Susana's husband is still renowned in the village for his extramarital affairs, but has never struck her or drunk too much. Their struggle, she says, was to feed and clothe their children. Most women speak of their life satisfactions in terms of their children. Susana (68) also speaks well of sex and the companionship of married life. Some women have always talked much with their husbands, others do not (in our sample, this is not a function of age). Some even have an ideal marriage, for example, Guadalupe (44) (see Chapter Eleven).

To some degree, for these women to talk of womanhood is to talk of suffering, enduring, coping, as it is for many other women in Mexico and Latin America. Suffering pervades their stories, partly because it has been their lives, partly because it is an affirmation of successful womanhood to suffer and perhaps partly because they see it as advantageous to project this image to the outside world through us.

WOMEN TOGETHER

Women talking

Women's relationships with each other are contradictory, both hostile and supportive. Even the possibility of friendship is in doubt. Not all women have women friends, or indeed anyone in whom they can confide. Many claim not to go out – as Victorina (33, Independencia) says firmly, 'The gossip almost comes to you in the house by itself.' Victorina confides only in her mother-in-law, who lives in the house:

> And that's because he says, 'Don't you talk to anyone, not even my mother! You bear it as you can!' [She laughs.] But there are times when you can't, you feel exploited, that – I don't know what. That day I found out [about the girl pregnant by her husband], look, I wanted to beat myself on the head or to go out, to – but then later I got myself in hand, and talked about it in the house.

To most women, visiting is not an acceptable activity. Imelda (55, Naranjales) says:

> Women don't visit. Take me, for instance. When my husband was alive he used to say, 'Never go around visiting because sometimes you say something and they go and say something else which isn't what you said – they get you into gossip, they get you into problems, so I never, never like you to go visiting.'
>
> So I got the habit of not going around. For instance, as I sell sweets, I sell bottled drinks, they come here – if they need anything, they come to buy.

[One neighbour visits] and now my granddaughter and my daughter, well, they are always here with me. That's it, but I don't [talk] with other people.

Rosario (34, Plan de Ayala) admits to talking to other women, but

In passing, like that. Because I don't have to go and visit my neighbours here, do I? I'm at work at home, from here to the stream to draw water, to wash, and when I come I'm back in the house again, or up there when we have a *milpa*, with plantains, but I – it really doesn't please me to go to other houses [. . .] There are women who, no, who really enjoy going around like that – visiting, and I don't [. . .] What I like is to do the laundry alone! [Several women said this.] Yes, yes, there's a little stream here, a spring – I go to wash there, where crowds of women don't go [. . .] because soon those are criticising that one, then that one is looking at another . . .

Clara (p. 174) really enjoys talking, and is allowed by her husband to do so, but she is in a minority.

WOMEN'S NEEDS

In our workshops and life histories, as we shall see in Chapter Six (pp. 125–6), women expressed both strategic and practical gender needs, but had far more constructive ideas about how practical rather than strategic needs should be met. That is, they had many ideas as to how to fulfil the role currently asked of them by society, but few as to how to improve women's position, how to resist what they do not want or how to change what it is to be a woman.

Individual women, when asked about women's needs in a community, commonly think at once of community needs and list the services which they do not have, whether it be water, electricity, a health worker, a dispensary (or medicines for the dispensary), a clinic, a telesecondary school (using television teaching), a shop, a local market, better access to the town, a park or basketball area for recreation or perhaps street drainage (as outsiders we notice that they rarely cite latrines). All the communities in which we worked had schools, but there was a frequent call for teachers to be made to spend enough time in the village during term for children to learn something. All these things are seen as women's needs, and to a degree an outsider can predict what women will say from a knowledge of what services exist in the village. So Aurora (53):

As a woman, what services do you see as necessary in Plan de Ayala?

Well, here, our children are left with the primary and we need a tele-secondary, because we want our children to learn, not to be left like this, like us, that they get ahead and escape from ignorance, that they shouldn't be like us, that they should learn more.

The other thing the women think of at once is work. Our very first respondent, Esther (32), when asked in 1990 in the very poor village of Laguna

118

Escondida what the village needed, said, 'We are doing fine. We don't need anything except work, and to get paid for it.' Sometimes they want work for women, more often for husbands and sons. In the poorer communities, women just want work; in the more prosperous ones, such as Independencia and El Tulipán, they want work for qualified people so that their children will not have to leave the area. In the richer half of our communities there is talk of industry, of an alternative to agriculture.

The other theme that is always raised is training, usually in income-generating activities such as making clothes, beauty care or embroidery, but sometimes in adult education or in the organising of groups. Training is problematic, for although enthusiasm is widespread, it conflicts with the requirement for women to be in the house, as Antonia (47) told us in Independencia. There, the State Department of Family Integration had sent trainers to work with the women:

> When the *promotora* [worker] came, yes, I went along. We had to, to get the benefits of the food handouts which they are still giving out. And yes, they give them to us, all for a small contribution – it isn't much, but it helps. They also gave us some materials to build latrines, one sack of cement, one of sand and six corrugated sheets – not enough to build it, that's sure, but yes, it helped [. . .] My daughter made a toy dog, and we made some bags too [. . .] and the women have learned, want to make progress. It's only that they say their husbands won't let them. That's a machismo that still isn't finished here, huh? Women have no right to an opinion – they don't let them fulfil themselves even though they want to. There are some who are intelligent, but their husband won't let them. If he tells them no, it's no.

Income-generation is also costly. Gloria (47) has views on what women need in El Tulipán, such as chicks to raise and sell:

> Well, for example, the consignment of chickens really helps, because at six months they are already laying – well cared for, they are laying already. Another thing I'd like would be a sewing class, which they say here they've been asking for, so that we could learn. Because one day I went to a neighbour who sews well [. . .] and I said, 'I want to see how you do the cutting out so that I can learn', and she said, 'Yes, come every day, but I'm going to charge you a little.' So I got angry and said, 'No, but many thanks all the same.'

Both the women and the men expect the state to buy them off with any kind of funds, the lack of which is a constant problem:

> *And do the nuns teach you to do anything that you can sell and earn something this way?*

> No. One of them taught us how to plant a kitchen garden. But you need money to get the soil ready, get the seeds and so on. You need money to do all that. They did it in school too, because the teacher got the seeds for the children. The idea was that the children would teach their parents. But how

119

are we going to spend time on that, and it might not even work? They also taught us some knitting. But you have to spend money on that too. I told her [the nun] once, 'You've got to buy the thread to do the little bootees and sell them.' But then we found the little bootees in a shop, much cheaper [. . .] It's difficult to sell here.

All these needs may be described as practical needs for the better fulfilment of their present roles as women. However, these needs are not only practical, for if many achieved substantial incomes, for instance, it might change what it is to be a woman, but that is rarely the objective expressed. The needs to control their own bodies and for safety from marital violence and rape are expressed in life histories and workshops and widely agreed on, but we shall see that women's groups see the strategic needs elaborated in this section as more problematic.

Women's groups

Women's groups are an important site for women to talk. The ideal woman stays at home, but there is also a very strong belief in the value of groups which is derived from national political and educational ideology. Women's groups, therefore, are both improper and highly fitting. Women traditionally meet in church, at prayer and while cleaning and decorating the church. Now, nuns in La Corregidora and El Tulipán offer classes on kitchen gardening and herbal medicine, while some other women may belong to sects and meet for religious discussion. Women who are Seventh Day Adventists prove highly articulate and confident, not because of the teachings of their religion but because of the experience of regular Saturday discussions.

The state in Mexico has traditionally promoted group enterprises. From 1974 to 1991, it encouraged each *ejido* to set aside a family-sized parcel of land for the landless women of the *ejido* which would become a Women's Agricultural and Industrial Unit (UAIM). Credit has, at least in theory, been available for groups of women who work these plots collectively. As Lourdes Arizpe and Carlota Botey pointed out (1987), agricultural work itself is problematic for many Mexican women and non-agricultural projects could be far more positive. However, a whole series of government programmes, aimed at reducing women's 'marginalisation', failed to analyse women's subordination and/or to introduce appropriate measures to reduce it (Robles *et al.* 1993). The women lack training and/or experience in group work, so that if a group is formed, it is often dominated by a clique which is able to monopolise all the training and credit, control activities and take all the profits.

La Corregidora, El Arroyo, El Tulipán and Independencia have such Women's Units but none strictly follow the rules. (La Planada has a unit, but it is only just being constituted.) Few women members (usually poor single mothers) actually work on the plot. Rather, in the Women's Units we visited most work is actually done by men, whether husbands or partners of the women members or labourers hired by the women to replace them. In Independencia, Victorina (33) actually withdrew from the unit because it had been decided that women were not allowed to get men to replace them to do the work. Groups need help in evaluating the commercial viability of their projects. There are training fads, identified from above, such as the killing and processing of pigs. In La Corregidora, the women have followed a two-day course on pig-processing but fear the investment in an unfamiliar activity. They want to start planting orange groves, for which they have much of the expertise. So far, their group has been in existence for ten years but the autocratic executive has (illegally) divided up the land to grow subsistence crops. In El Arroyo (Chapter Nine, p. 163), a clique commandeered the credit to raise cattle on the land, and in Independencia one (Protestant) family is in command. Women officials in local offices of the Ministry of Agriculture or of Agrarian Reform are charged with the training and support of these groups, but very rarely go to the communities even for an hour or two. These officers are important in brokering finance, but make very little other input and it is difficult for women who travel from the communities to the offices to find them. We often couldn't find the women officials although we stayed in the town and called again and again over a period of days.

Yet, the women see these units as a positive force for all kinds of organisation. To the outsider, it may seem that they occupy women's energies with little hope of profit, but to the insider, they are women's meeting-place and strength. The presidents of two of the groups could neither read nor write, but travelled to meetings in distant towns and cities. Many women are extremely critical of the corruption and inequities, but still believe in the ideal, in the possibility of earning an income through such a group, if only they could be trained. Isabel (51) has been president of the La Corregidora group for ten years and faces rebellion from other members, but her ideas are positive:

You as a woman, what services do you see as necessary here in this community?

I've talked about my ideas to colleagues in the UAIM, that it's our duty. Because we have to do something for the village, to work with the people to clean the streets, to make a kindergarten [. . .] Many people think the government is going to come and give us things, but that isn't so. And I, yes, my idea is that we should join up, we should make ourselves like block leaders, so that one should take charge of one block and another another, to get things cleaned up, so that the men should look at what women can do [. . .] If women start to do something, men criticise – they don't work or let us work, that's the trouble.

Strategic gender needs are discussed primarily in the context of women's groups, and prove to be very much those identified by urban, Western feminists: women's rights not to be battered, not to be raped by their partners, to control their own bodies and fertility, to dignity and respect. Solutions are seen in terms of 'education' and controls on the purchase of alcoholic drink, but, when women are asked, 'How can we solve this problem in our community?' they do also discuss what could be done within the group.

CONCLUSION

Our aim had been to ask pioneer women what they saw as their problems and what solutions they proposed, expecting in our ignorance that they would focus on the difficulties of pioneering, of establishing and maintaining a livelihood where there had been forest. Ours was still a socialist-feminist agenda, concerned with production and perhaps services, and certainly these were the issues we expected pioneer women to be willing to discuss with outsiders. When, after the questionnaire or in the course of life stories, we asked, 'please tell me about women's problems in this community' and 'about your problems', most women indeed replied at the household level and listed the practical needs discussed in the previous section. Some would also speak of alcoholism. It was when women were in groups that similar questions would also elicit responses that included the need for change in gender relations, around conventional, Western, urban, feminist issues which we had not expected them to raise. These issues were also important in life stories, but very rarely expressed in response to questions about women's problems. The alternative context of group talk or of developing a life story was critical in women choosing to speak of what they see as personal, private concerns.

6

OUTSIDERS' CONCLUSIONS

OUTSIDERS

To the outsider, it appears that there are three components to women's difficulties in land settlement. First, '*The lot of women is often worsened by settlement*' (Chambers 1969: 174, emphasis added). However, this may not always be the case, for Susie Jacobs (1989) thinks that the lot of women in Zimbabwe has been improved a little by land settlement. But, very often, women lose rights to land, access to income and social support while acquiring a greater workload (Chambers 1969: 174–5). This happens because an international culture has developed around the planning, administration and legislation for land settlement which has been curiously unaware of social issues and has brought radical and unintended changes in the relationships between women and men. This planning culture has proved very resistant to both reason and experience. Thayer Scudder's long campaign for attention to social considerations (1969, 1981, 1985, 1991) is bearing a little fruit at the World Bank but, for the lot of women in land settlement in general to improve, the international culture around it will have to change fundamentally.

Second, *many land settlement schemes lead to poverty rather than prosperity,* being promoted by the interests of politicians, bureaucrats, aid agencies and the private sector rather than by any cost-effectiveness in increasing production or livelihoods. The harvest of this dominance of political interests in Colombia is bitter poverty (Chapter Three), as is the lot of millions in land settlement around the world. Many feminists believe (Boulding 1983) that poverty affects women most severely, since women have least access to household resources. Most poor women in poorer countries see poverty as the great enemy and we, as feminists, follow Gita Sen and Caren Grown (1987: 80), for 'We want a world where inequality based on class, gender and race is absent from every country.' Again, very dramatic change is needed if land settlement is to be promoted only when settlers will actually benefit on a sustainable basis.

Third, particularly in Latin America, *much settlement is not sustainable*: livelihoods disappear and pioneers move on to 'open up' more forests. For the welfare of the people and the protection of biodiversity, feminists and conservationists both see sustainable livelihoods (p. 26) as the solution. Conservation is possible only if people have a better alternative. In the Magdalena Medio in the 1960s (Chapter Three), the Corporation of the Magdalena Valley sought to control overfishing and deforestation with armed guards. Naturally, this action saved no fish or forest but gave conservation an extremely bad name with rural people, as the prosperous were able to bribe the guards, while the poor lost their livelihoods (Townsend 1976). To better the lives of poor women and to protect resources for the future, feminists and conservationists would agree that it is vital that the uses of land already settled and the lives of people there improve.

We designed this research in order to formulate guidelines for planners to enable women to join in and gain from land settlement. The questions that need to be answered are how far are there general lessons that might enable women to better their lives? And, are all the answers specific to countries or even to places?

Outsiders and the Mexican case

Some of the problems faced by the Mexican pioneers would be very familiar elsewhere. The very isolation of the frontier means that remote communities are at a big disadvantage both in the market and in getting services from the state. However, many Mexican problems are specific to Mexico, for government policies, past and present, have created very singular conditions. Poor rural Mexicans have for long been manipulated with minor palliatives – never solutions – and have developed a highly dependent attitude to a state which is now withdrawing many of the palliatives and asking them to 'take responsibility for themselves'. Within this Mexican system, there is great diversity among communities. In Los Tuxtlas, Palenque and El Tulipán, climate, soils and access are suitable for the intensive production of beef and milk. But the World Bank's 'market-friendly development' (1991) is not happening here and markets, credit and training are simply not adequate to the creation of profitable, let alone sustainable production. More positively, the future of the Oxfam-Belgium project in El Tulipán (p. 7) may be important, for it concentrates not only on techniques but on working in groups, on doing accounts and on planning. If its success continues it will be very significant, as may agroforestry research in the Sierra de Santa Marta, Los Tuxtlas. There are possible futures for these communities – they need not all decline into renting their land to graziers, or even, under the laws of 1991 (see p. 52), selling it. They may find alternatives developed by themselves, by Oxfam or by researchers.

Within this diversity of pioneer problems, many of women's problems are directly linked to pioneering and to the changing relations with the environment. The newness of the communities contributes to the lack of services and, through women's separation from the support of their families, to men's freedom to drink, batter and rape. Changes in the agrarian system also deprive many uneducated men of their pride in providing for their families and create conflicts around women's fertility, adding to the violence, rape and alcoholism.

In August 1991, at the end of our time in these communities, we, the authors, sat down together in Escárcega, Campeche, to set out an ordered structure of proposals which might benefit pioneer women in Mexico, beginning with the community level and ending with the national. We did not take our proposals back to the communities to consult, but the structure is informed, we hope, by what we had learned. Our proposals, which now follow, begin at the community level, for individuals have already acted and some women have achieved prosperity, while others have left violent partners. What immediate actions could bring positive change for more than individuals? We ask this in Caroline Moser's terms (1993; see p. 49): what might help women to meet their practical and strategic gender needs? All are eager for success as women (Chapter Five), caring better for their families, offering more to their children, fulfilling their roles more successfully. None wants to escape these 'women's roles', but many also want strategic changes in their relations with men, better access to income (also a practical need), respect, autonomy and freedom from fear.

Action at the community level

We do not see the settlements we studied as ideal (mythical?) communities where all are equal, loving and co-operative, but the community does provide a real framework. Most communities in rural Mexico are run by an elected committee, usually all male. Community organisation is more formal in *ejidos*, where all members meet at regular intervals, but *colonias* also elect officials. In El Arroyo, the elected leader, Adolfina, has an impressive record of using her femininity to bring in state help, although much of it has been wasted through community feuds. Elected officials or community votes can achieve a great deal in meeting women's *practical needs*:

- Mexican women depend on the ability and willingness of community leaders, men or women, to extract public services from the municipality and the state, which accounts for the variety in the level of public services. Many community leaders could do far more. Women could also co-operate more often to bring pressure on leaders and help them make demands.
- Community priorities are often strongly male-oriented, failing to take into account the very different interests of men and women. In Tacaná, women's prime concern is wells for water, since they must carry all water a kilometre, but men's prime concern is the poor access to market. The community has done all it can with its own labour, but equipment to improve the track or drill deeper wells must be sought from outside. The priority of the community leaders, all men, is the road. In such cases, women could make more direct collective approaches to outside authority, as they did in La Corregidora and in Tacaná itself, using us as secretaries.

125

- Women often complain bitterly that schoolteachers, as in many other poor and remote areas of Mexico, appear irregularly and infrequently in the village, so that children may attend school only a few days a month for many years without learning to read. Communities could demand sanctions against such teachers.

- Men expect to contribute labour to community projects which are often the concern of both men and women. Men were building a classroom in Plan de Ayala, digging to lay piped water in El Tulipán and had just built the track to Tacaná. Sometimes the state provides materials and the community labour, as for the DIF centre in El Arroyo (p. 68). Many of women's priorities could be met in this way, although it is important to avoid the excessive demands we met in La Villa (p. 177).

- Each community has a health committee, for instance, which can mobilise community labour for projects ranging from street cleaning to water-supply. This committee has considerable power to raise awareness of hygiene and to organise health education, but while some are excellent, others could do much more.

- The community can create public spaces. In the more prosperous places, many women call for a regular market to be organised. Everywhere, women propose a basketball pitch or a square where families can walk and mix in the evening as important supports to family life and alternatives to the bars.

The community could meet some of women's *strategic needs*:

- The community can exert pressure on the municipality to close some or all bars. Women in El Arroyo and Jasso have achieved this and say that drunkenness is reduced, although not eliminated. (Some men go out of the village to drink, and may spend more and stay longer, while others bring drink in.) As we have just seen, the women see sports pitches and squares as offering alternatives to bars.

- Community authorities could act much more often and more strongly against violence towards women, even within the present law, without waiting until murder is committed.

Women actually see these as practical needs, as ways of living up to existing ideals of womanhood, while we see them as strategic needs, as men's sanctioned violence is so fundamental to relations between women and men in Mexico. There are many actions available to community leaders to improve women's lives, and in some cases women have forced such action. However, if the government's efforts to 'privatise' these communities by its changes to the national constitution are effective, many women will lose their minimal economic security (p. 52), but all will lose their main way of getting things done in the community.

Action by women's groups in the community

A message has reached these women from outside, saying that they must act together. Traditionally, this is both acceptable and unacceptable (p. 120), for women have grown up with talk of collective action, but collective work has been for men. Now, women are responding to government encouragement and international trends by seeing new possibilities for themselves if they work in groups, hoping to find in them the strength to deal with the market, the bureaucracy and their men. The principal forum for women is the UAIM. Religious groups, providing the other main opportunities, seem far less significant, although almost all women are believers. Seventh Day Adventist women demonstrate the potential of women's groups by being often the most confident and forthcoming in the village, as if the Saturday women's meetings to discuss the Bible develop women's ability to express themselves and indeed their self-images (see Chapter Eight). Both the Catholic majority and the abundant evangelical sects encourage women to accept their role in life but offer little practical support. Women's groups exist chiefly to clean the church or chapel and organise festivals, and rarely have even a charitable function, let alone offering possibilities for any exchange of skills in, say, sewing or literacy. In Sor Juana, Liberation Theology promotes agricultural training among the men, but does nothing with the women. However, women do very warmly welcome the banning of alcohol by evangelical and Adventist groups (p. 152).

Women rarely meet as a group unless there is a UAIM (p. 120), legally required to work collectively. (The constitutional changes are vague about UAIMs and include no proposal to replace them.) Internal divisions and fraud are common in UAIMs as women have no experience of collective work and no training. We feel that UAIMs are a political device, giving women a little land on restrictive terms and channelling their energies with scant hope of profit. With Lourdes Arizpe and Carlota Botey (1987), we would prefer to see more real, appropriate opportunities for rural women in Mexico. Nevertheless, these groups have clearly become the focus for women's concerns and are the institution which offers the most prospects for women's pursuit of their *practical* and *strategic* needs. We think groups have real potential to help women achieve their goals, whether concerning wife-battering, income or training, but that women are severely hampered by their isolation and their inexperience in group work.

Action by community workers

What pioneer women in Mexico say they want from the outside world is 'training'. New problems confront them in the economic sphere and they see much of the solution in education and in the acquisition of new skills. Their talk is of training, of self-improvement, of progress, of participation in the public sphere. These urban, North Atlantic, 'modernising' ideas had arrived before us.

In our view, there is a great deal of 'Women in Development' work in rural

Mexico, nearly all oriented to *practical* needs and tending to use women as tools to increase production and, in the face of national crisis and growing poverty, to achieve better welfare with fewer resources. Not only do the new tasks painfully expand women's workload, but they often require a week's work to yield the equivalent of a day's agricultural wage. Women community workers (outsiders), usually with minimal training, are very important in introducing these programmes. Very few reach remote communities, but there had been such workers in Independencia (now very accessible!) and one in particular had been welcomed with enthusiasm (p. 119). She had, unusually, sought to raise women's self-esteem, but had not enabled any discussion of conflict, whether about domestic violence or alcoholism.

We think that the minor skills being offered to help women earn more income are an inadequate palliative, and that, for a worthwhile return on women's time, far more and better-trained instructors and different programmes are needed. Women showed us that they have ideas and knowledge for potential projects, but need help in evaluating and implementing them, particularly in accounting, marketing and, where applicable, group work. Non-traditional projects will be essential (Arizpe and Botey 1987). There is appropriate experience in Mexico: Emma Zapata (1990) has worked with rural women's groups in the highlands, developing training which shifts the emphasis from using women as tools for the good of others to enabling them to achieve dignity and autonomy while working for their own objectives. This training very closely parallels that which women pioneers say they want.

In all our workshops and in many interviews and life stories, women spoke of drink and violence as great, ever-present, intractable problems. For many, these problems must be 'solved' by the woman herself, but group action is always mentioned as desirable, either to protect the woman or to shame the man. A group, they say, could agree that a woman could take her children to another house when her man comes home drunk, or 'all members could go to the house and sit on him' (workshop, El Arroyo). We did not observe successful solutions being developed, but groups at least can give an opportunity to confront these strategic needs.

Our main conclusion is that large numbers of high-grade workers, whether from the state or non-government organisations, could do much to enable pioneer women to pursue their objectives, individually or in groups. Women are traditionally isolated, but many now desperately want to find out how to earn, and see groups as providing a good framework; possibly in response to urban changes and the media, women also want to resist violence and alcoholism. Good community workers could offer a great deal in helping women to work for what women want and to escape from the traditional dependency on the state which controls the Mexican rural poor so strongly.

It may seem strange that we recommend work by outsiders. In part, we are responding to a discourse in which politicians have talked for decades about outside support for local, collective initiatives. In part, we are saying that women do see group work as an objective, but lack relevant experience, particularly with accounts and organisation. Conditions are very much against their achieving their goals without training, which is simply not locally available.

Action by non-governmental organisations (NGOs)

Again, pioneer women are eager for what Mexican NGOs can offer, particularly the skilled community workers we have just described, and many also long for access to specialist groups such as Alcoholics Anonymous. The NGOs work mainly in the cities and accessible rural areas but could achieve much in these remote communities. We do not see NGOs as a general panacea, but, at the moment, many in Mexico are extremely effective in offering just what pioneer women seek to extract from the outside world.

Action by the private sector

In Los Tuxtlas, pioneers developed skills in growing imitation rainforests (p. 61) which achieved the objective of much expensive research across the tropics by producing a great range of products with very few inputs, since their biodiversity reduces the need for fertilisers and insecticides. This skill is being lost as ranching takes over, but, given private sector investment in centres to process the products according to season – no small requirement, and many are seasonal – we think that national and international markets could be supplied with exotic 'organic' fruit juices, fruit yoghurts, ice-creams, teas and timber. Fruits include pineapple, avocado, lime, orange, grapefruit, mango, watermelon, melon, guava and, used more to make drinks, the less familiar nispero and zapote. Herbal teas from Los Tuxtlas are already marketed in the United States. Such an approach to national and international markets is, we think, the one way in which these pioneers can offer a high skill. It is also important to note that, although monocultures are far easier to process and market, they are not sustainable. Although much contract production in poor countries is highly exploitative, it is an alternative to the massive loss of jobs and the high level of complex skills could provide some defence against unemployment.

Private capital may also eventually turn its attention to women's enthusiasm for homeworking, as it has already in many other parts of Mexico (Wilson 1991; Benería and Roldán 1987). Unfortunately, the only attraction pioneer women can offer is the extremely low level of pay they would be prepared to accept. Many women would welcome the chance, but experience elsewhere suggests that they would work very hard for little financial gain even compared to the agricultural wage.

Action by the state: a view from below

In Mexico the municipal, the regional and national bureaucracies are charged with responding to requests from communities, but the massive cuts in state rural expenditure now make this duty very difficult even to attempt; so does the bureaucratic culture in which achievement and incentives are centred on urban offices rather than rural communities. We think that there are still specific needs which could be met cheaply, often by implementing existing rules. The municipality can act with respect to strategic needs in the cases of violence and alcoholism, as many have by banning the sale of alcohol. No full evaluation of such bans has been made, but Greenberg (1989) found that murders had dropped dramatically in one community which banned alcohol and we found that many women were positive about the experiences of Jasso and El Arroyo. None of our workshops, interviews or life stories produced any proposals for action by authorities against wife-battering: the idea is a joke, for authority is 'male'.

If we may dream of Utopia, we suggest that national ministries could meet many practical needs by following their existing rules. The Ministry of Education could require teachers to fulfil their contracts (as it already does in some areas). True, the communities are remote, living conditions poor and salaries very low, but most pioneer families are even poorer than the teachers they so desperately need. The Ministry could impose sanctions on absent teachers and provide many fewer mysterious excuses for teachers to 'go to town to report'. The Ministry of Health could enforce its instructions among its own workers with strategic effect. Too many women's lives are at risk because their husbands can refuse them contraception and then rape them (p. 107). Too many women lament having been sterilised against their will (p. 107). Such actions reflect not the law but the local practice of clinics. Basic primary health care is also badly needed but will be expensive.

We found support for UAIMs from the Ministries of Agriculture and Agrarian Reform to be token and office-bound. More *ejidos* could have UAIMs, and trainers could be required to spend months in a community, but there are no incentives to do so. Living in remote communities would be hard for these government workers, who have struggled to reach these posts and have no desire actually to be in the communities or even visit them. However, incentives could be created (or the posts abolished).

The DIF (the Department for Family Integration) has great scope, we think, for effective but costly expansion, especially in its provision of training for women and the young which is still largely a palliative. Many of its women workers are young and single, and could be sent to more remote communities so long as they speak the language. As with officials in agriculture, only a change of culture in which rural communities, not urban offices, become the centre of their work could justify the employment of these workers at the taxpayers' expense. Such promoters could be asked to pay less attention to women's domestic tasks and minor earnings and more to their rights and real earning potential. In strategic terms, the DIF could do a great deal to encourage women to resist battering and to develop support from the community and the law. Adult education has also much to offer, again at extra expense. There is

130

opposition from men, but many women see it as vital to practical and strategic needs.

Action needed at the national level

The 'solutions' we have proposed are all short-term. The national government could provide, in the long term:

- effective rights to battered women,
- more support to rural areas,
- less support to extensive ranching,
- a new policy on colonisation,
- rights in the farm for 'farmers' wives' (reduced by recent constitutional change), and
- much more efficient public services, etc.

These decisions are a matter of national policy, to which distant voices from the rainforests do not, but should, contribute. In writing this book, as Mexicans and a foreigner, we are saying that there is a lack of effective democracy in Mexico.

INSIDERS

Who are the experts? The prescriptions just discussed were created by us, however hard we tried to listen. We have been extracting issues, the etic, while pioneer women think of their whole lives, the emic. We personify different cultures, experiences, priorities as well as (the strength of the etic view) bringing experience from other parts of the world, ranging from the production of 'organic' fruit to the pressure from the United Nations Commission on the Status of Women for more action on the issue of more violence against women. We agree far more with each other over 'solutions' than with our subjects.

It seems to us that the insiders, women pioneers, would at this moment welcome 'development' of any kind and hope to turn it to their advantage, that women would accept almost any exploitation of their men, their children or themselves which led to increased income. In the current crises, both in the national economy and the local agrarian system, women rank additional income over sanitation or clean water or reduction in domestic violence – items which we might rank more highly. Their expressed wants are for their families, not for themselves. Pioneer women ascribe almost no value to their time and will work all week embroidering when we might prefer to dig in the sun for a day for the same pay. Pioneer women do not see the commercial potential of which we dream in their garden skills, but speak with longing of new forms of employment elsewhere in Mexico, such as packing plants for monocultures (as in the strawberry industry), homeworking or contract production (p. 71).

Who are we to disagree? Many Mexican women take a positive view of these activities. For example, women and men in Santiago Tangamandapio,

Michoacán, see the homeworking they have built as a great achievement, not as oppression or exploitation (Wilson 1991). What authority have we to judge? Our frames of reference are different from those of women pioneers but not necessarily better for them. David Lehmann once said to a conference audience, 'Avoid the concrete peasant', which tends to reduce us to 'experts'. In 'development', we have not begun to exhaust personal testimony, to listening to the emic.

We outsiders are aware of positive action by the state in other countries. Women in the communities we studied are extremely critical of local community leaders, the municipality, the state and the national government, but do not see these weaknesses as capable of being solved or even reduced. Women speak of them as uncontrollable, like drought: one does not think how to change the weather, but how best to plan for or respond to it. Most women have an extremely low opinion of the public sector, although they depend on it heavily for services.

There is an extraordinary enthusiasm for women's groups, considering how isolated many have been and how strong are the conflicts of interest between the successful and the near destitute. Most want more collective action and are eager for training. Despite all the diversity and difference between pioneer women, this is a moment when most are eager to co-operate and when many men, conscious of the immediate needs for jobs and services, support them. Pioneer women have great faith in education as the answer to everything: to the lack of jobs, to conflict in groups, to drunkenness, wife-battering, etc. We do not entirely share this faith in 'education', and such disagreements are one problem of 'advocacy research', of wishing to represent the views of others (Townsend 1994). We have tried to be as clear, as transparent as possible about our role, but at best we can only seek to interpret what we think we heard and to identify our differences of opinion.

WOMEN IN LAND SETTLEMENT

What are the lessons of these considerations, this experience, for women in land settlement worldwide? In general, we hope that land settlement itself will become less popular with governments and aid agencies. Where it already exists, and where it does offer better returns than investment in long-settled areas, we want to see its whole culture change. We agree with Thayer Scudder (1991) that a wider social science perspective is essential, but we are concerned about the constant search for general rules, general solutions rather than attention to the national, the regional, the local. Of course 'the global' is relevant, for international debt is one force behind land settlement. But a decent life for women in many pioneer areas of Colombia would depend on both solutions to the national problem of drug trafficking and new power relations between men and women pioneers. Possible solutions to the problems of Mexican pioneer women again range from the local to the national. It is hardly believable that positive change would not be needed at all levels before Mexican pioneer women could solve many of their problems.

Women's loss of rights to land, access to income and social support, while their workload is increased, is a very common but not universal outcome of land settlement. From the literature and from our work it is clear that improvements are likely to be specific to countries, cultures, even places. The planners and managers of land settlement certainly need a better education in social science but even more in the cultures involved. They also need to be less autocratic and managerial, more interactive, more responsive. Time and again they have imposed inappropriate models with disastrous results and may well be encouraged, unfortunately, to do so again (Hulme 1987). As outsiders, we do not argue that women's identification of and solutions to their problems present a whole answer, but that theirs is a very important voice, usually completely unheard. 'Participation' is all too often 1990s-speak for 'Do as you're told – or else!' but, at its best, participatory appraisal can be the most appropriate education for planners, managers and administrators (Chambers 1992).

OUR CONCLUSIONS

We call, then, for more responsive, locally informed, participatory planning. In the long run, we should like to go further and call for an ear in 'development studies' to the voices of individuals. 'Development studies', like 'geography' (Rose 1993) seems to us to have a thoroughly male gaze, generalising, ordering, objectifying. Mexican women, in workshops and through their life stories, taught the authors of this book that even as women researchers we had come with a male agenda, insensitive to interpersonal issues. We have detailed our proposals for the needs of Mexican women as we outsiders see them, and have tried to show how these proposals differ from pioneer women's own solutions. We have argued that solutions must be local and specific.

We want now to go further. Much land settlement follows inappropriate international models with little regard for local social or economic reality. But not even local reality is homogeneous, and individual pioneer women in Mexico think very differently about their lives. We want now to demonstrate this, to give space to the voices of four women, still edited and constrained, but with more scope to express a little of their character and individuality. 'Standpoint theories show how to move from *including* others' lives and thoughts in research and scholarly projects to *starting from* their lives to ask research questions' (Harding 1991, emphasis in original).

We invite readers to start from some lives of pioneer women as, in our next project, we hope to do. In Chapter Five, we used quotations from pioneer women to describe their lives, but this can only do violence to what they told us, because we cut it up, use it out of context. Now we present four voices. We regret that we retain control over the text (Davies 1992). However, we present the voices not as commodities for consumption but as experts in their own lives.

Part 3

INSIDERS' VOICES?
MEXICAN WOMEN
SPEAK

7

RE-PRESENTING VOICES
What's wrong with our life histories?

Susan F. Frenk

When Janet first asked me if I would like to collaborate in this project by translating and editing some of the life histories of the women who had been interviewed, I was both excited and anxious. The prospect of engaging in a group effort to carry out research that would be as far as possible driven by the needs and desires of the women interviewees and to negotiate with them as between subjects, rather than subject–object, seemed to offer a chance to carry through some of the aspirations that our feminist research group had been articulating over several years of lively, sometimes anguished, discussions. On the other hand, those same discussions, along with the reading and thinking in cultural studies generally in which I had been engrossed, made me acutely aware of the dilemmas of the translation process. For the past two decades, and the 1980s in particular, have been a fertile period for thinking about communication generally, in which a view has emerged that all 'communication' between people is not a transmission of pieces of information but a complex process of interpretations, of shades of meaning. Our original interviews were one such multifaceted process; 'communicating' to outsiders is a second; translation studies have shown that 'translating' them is a third.

Notions about translation have their own histories in our own and other cultures. These are elaborated by recent research (Basnett and Lefevere 1990), which we cannot explore fully here. In the UK, however, although there is some jostling for position between models motivated by linguistics and those connecting with cultural studies, all engage to some extent with the idea that the specific cultural context goes far to determine the meaning of what is said and heard. This can be seen as a spectrum of webs of associations. A relatively straightforward, easily identified example is that in the UK we use the same word, 'hello', as a greeting when we meet people and when we answer the phone. The French, however, use *allo*! ('hello') to answer the phone but will greet people they know with 'ça va?' ('how are things?') and people they are meeting for the first time with *enchanté(e)* ('delighted'). A good dictionary should indicate the different contexts in which different phrases are appropriate. The example does not stop there, of course, because there are variations in the greetings used by people who differ in age, gender, social class, ethnicity, region and so on within France. Some of these variations reproduce conventional relationships of power, status and identity, such as the idea that an adult should be addressed in the *vous* (formal 'you') form in some contexts (meeting someone for the first time, for example), whereas a child may be addressed immediately in the *tu* (informal 'you') voice. As recently as the 1960s, even some adult French people would still have addressed their own parents in the *vous* form. Changing social relationships, including family relationships, have produced a loosening of this particular convention, but it had already been violated by young adults who choose to define themselves differently from their parents by using the *tu* form among themselves from the first meeting, notably in the student movements of the 1960s. What their parents would experience as a lack of respect, the students perceived as a democratic act. For an English-speaking person, the nuances of the shifting boundaries of *vous* and *tu* have no direct linguistic equivalent; nor do the webs of relationships in which they are enmeshed (Basnett-McGuire 1991). Even here, then, in the apparently simple world of everyday 'communications', with an apparent 'equivalence' of context, where it seems to be a question of finding the appropriate phrase for a 'matching' situation, we can see that the differences may actually relate to different ways of experiencing the world. Both within one supposedly unitary 'culture' like France and between cultures, there are different ways of understanding relationships such as 'family' and 'friendship'. This is easier to grasp with an idea like 'democracy' which we have learned to think of as more abstract and in which we can see competing definitions being bandied to and fro by politicians from different parties or nations. It is harder to apply it to everyday experiences which we take for granted, or emotions and relationships which we think of as 'natural', and assume to be experienced in the same way by everyone, everywhere.

For some, this recognition has led to a position identified with the work of Jacques Derrida, that every 'translation' creates a new text rather than a version or copy of another text. In other words, the translator moves from one 'original' to another. Underlying Derrida's writing on translation is a complex philosophy of language and meaning which has come to be known variously as 'deconstruction' or 'post-structuralism'. Space does not allow us to discuss it fully here, or the important critiques made by feminists and other critics, but I would like to bring into focus the idea that every utterance in our interviews would be 'untranslatable' since it is shaped, framed, configured by and for both its immediate, contingent contexts and its cultural context, for which there is no equivalent in another language.

This raises critical issues in cultural politics. A language, in simplified terms, expresses what its powerful speakers, such as men, want said. Women and other subaltern groups have a different relationship to their language (Cameron 1990) and to what can be said. Such subtle nuances are readily lost, distorted or exaggerated when the utterance is translated, reconfigured, rearticulated.

In recent years, there have been developments both in Latin American studies and in an area that has become known as postcolonial studies – although one of the figures most identified with it disputes the term (Spivak 1991) – which offer possible routes out of the impasse of untranslatability and cultural loss. In Latin American cultural studies, the concept of 'transculturation' requires 'translation' to include the cultural context and to explore a more dynamic model. In that process the two language-contexts simultaneously transform one another, so that the loss incurred is not absolute, even in situations of unequal power such as colonisation. Some of this work focuses on bilingual, bicultural texts which mediate the two contexts (Rama 1982); some of it concentrates on what is, and has been, written or spoken by subaltern groups (Basso 1990); a third strand also analyses the texts of the colonising or dominant groups (Pratt 1992). In this way, the traces of the less powerful may be sought in a range of resistances: in counter images of themselves, in their own refashioning of dominant discourses, and in the gaps and contradictions, in the very production of new discourses by the powerful. In the life histories in this book, we can see women refashioning dominant notions of motherhood, for example. We have tried to take one step towards transculturation by preceding our life histories with accounts of the environments and societies in which the women live, in Chapters Four and Five.

From the space of contemporary South Asia, Gayatri Chakravorty Spivak has responded to Derrida, via both feminist and Marxist modes of analysis, to counter the philosophy that occasions are so unique that there can be no comparison across time or space, no real history or geography and no collective politics. The powerful create images, representations, of subaltern groups such as women, children, the poor, ethnic minorities, as an integral part of the creation of their own self-images and their own justifications and 'explanations' for their power. Gayatri Spivak's project is simultaneously to deconstruct (to break down and expose) these representations and to rescue the cultural difference of these subaltern groups, their identity, by displaying how this difference is produced and reproduced in a double process: both in the positioning of the dominant groups and in the counterpositioning of the subaltern, from the margins.

The current dominant practice in translation is simply to present a finished product, made to look 'natural' in the final language and to conceal all the difficulties. Translation as *transculturation* or deconstruction requires the translator to describe carefully the process of translation and rearticulation (Basset-McGuire 1991). A new set of dilemmas follows: for instance, will the reader be able to understand it? Spivak's own work is immensely challenging even for readers well versed in the various forms of feminism, Marxism and deconstruction which she deploys. If one of the aims of this book is to make the voices of the women who gave their life histories to the project accessible to a wide-ranging English-speaking audience, should we present the 'finished product', concealing the difficulties, or set out to develop new narrative forms? If we do the latter, to how many readers will they really be accessible?

The life history as written autobiographical narrative has increasingly been the focus of study in terms of the different ideas of the self which it has embodied in different periods in Western cultures. Recent work in Latin American cultural studies traces the ways in which Latin American forms of experience reshaped the conventions imported from other cultural contexts (Molloy 1993), but equally explores the constraints of narrative conventions, social convention and self-censorship which operate in this kind of writing. Many of these issues apply to the life history as oral interview. It is not merely a matter of avoiding leading questions (this has received considerable attention in the social sciences), but of more subtle ways in which the interviewer may be partially determining the responses. How do we avoid a line of questioning resting on implicit assumptions that are culturally determined but seem natural to the interviewer, as opposed to questions in which the moral agenda is more readily apparent?

Most of us were inexperienced in taking life histories; with hindsight, we should have begun with more formal training, and transcriptions should have been checked more thoroughly during the fieldwork. Leading questions are sometimes necessary, but indefensible leading questions did find their way into interviews: 'What do you suggest, something to improve the situation of women in this *ejido*? A school or something?' (see p. 152). We can now identify these with some embarrassment, although the women in many cases answered against the grain of the question, which suggests that the situation established in the interviews was sometimes sufficiently flexible to counter some of the problems. We need to spend more time studying the micro processes of these oral exchanges, to understand the many possible dynamic relationships of power, collusion and silence which can arise (Kedar 1987).

The issue of the less visible ways in which culturally specific narratives shape the life stories is again such a vast field that it cannot be covered adequately here, but there are accessible texts for readers who are interested in pursuing it (Stanley and Morgan 1993; Smith and Watson 1993). Questions we now ask ourselves include: have we elicited a vision of the women's lives heavily biased towards their childhoods (as they depict them now, from the present) because we ordered the interview in that way? Would the picture have changed if we could have returned for a further interview which began at a different point? Are the questions about how they envisage their future after the children have grown up, or if they were to lose their male partner, formulated too much from ways of talking about and projecting, organising, our lives in our own contexts? Has it produced a particular ending – and future – to the stories that would not otherwise be there?

At the point of translation the issues surfaced again. When the women talk about sterilisation they tend to use a passive construction, 'he sent me to be operated on', 'they operated on me', 'he had me operated on'. How far can this be attributed to the general use of such constructions in Spanish? How far can it be interpreted as expressing a situation in which the women had little, or no, power in the decision? To translate it as 'I had the operation' reads more 'naturally' in English, but it conceals the ambiguity of the Spanish phrase. A careful study of language use in the local contexts would be necessary for a more confident interpretation.

When we come to translate the life stories a further issue arises: we are changing an oral text, in this case a sort of uneven conversation, into a written text. There were two models available to us which would be familiar to our readers: interviews in magazines and dialogue in novels. We have chosen the magazine format because it maintains a greater sense of the intervention of the interviewer, even though the life histories were designed to allow greater power for the person being interviewed to digress and to redirect the dialogue. The problem with this model is that it loses the inflections of voice, facial expression and gesture of the original conversation, and these all affect the interpretation of what is said. Of course, television programmes and videos are equally open to misinterpretation, since the meanings of the whole repertoire of the body are both culturally shaped and ultimately open to the subtle combinations through which we come to identify individuals. However, at least a fuller image is there to be interpreted, whereas our current models for transcribing oral exchanges are woefully limited (Basso 1990). These problems in representing 'voices' extend to the patterning of speech, both recognisibly drawing on conventional language patterns, yet individuated in the combination of variation and voice. Phrases that punctuate and shape the flow of speech are markers of social class and region, in different ways, in both cultures. The closest approximations in English might well be dialects, which would have to be translated again by at least some readers. In the end, we compromised between the language that I would use myself in informal and formal situations and language drawn from friends of different backgrounds and experiences where I felt it offered a better attempt at transculturation. We now discover that some of the phrasing is unfamiliar to North Americans, because the friends on whom I drew had all been British. I hope that this unfamiliarity will remind readers of the diversity of dialects in Mexican Spanish. A further difficulty is that the repetition – of certain words and phrases – integral to oral communication is often edited out in published interviews because it is considered 'boring' to readers; yet it is a crucial part of the 'voice'. This can extend to other kinds of repetition, such as recurring topics or opinions, yet the fact that they come up again and again, sometimes as digressions from the current topic, suggests that they are impor-tant. Janet and I realised that we had edited out a considerable amount of such repetition when women were talking about their health worries. Ironically, we can now see that this is one of the very few topics which permit the women to speak about needs and desires as a more separate 'self'; mostly when they are invited to talk about themselves the focus inexorably shifts to children, to spouse, to a self always seen in relation to others. This is not to make a value judgement, since different notions of the self and the kinds of social relationships they imply, or construct, are hotly debated by contemporary feminists, but it does seem to contrast with our own experiences of dominant notions of the self in the West, where we are encouraged to think of ourselves as separate beings, with individual needs and desires to which we have a 'right'. It may be a mark of cultural difference, therefore, and if there are different, indeed conflicting, notions of the self at play here, we may have erased its most insistent moments! Yet we were both uneasy about translating the way in which the women talked about their health and this partly explains the cuts we made. Time and again they talked about 'nerves', something which did not appear to be denigrated by the doctors in Mexico whom they consulted but which has all kinds of negative connotations in the UK, despite attempts by feminists to popularise the results of sympathetic studies of women and health. In Britain, a woman who has 'trouble with her nerves' is often considered unreliable and her words and behaviour are received with a certain scepticism. We did not want to risk this unconscious reaction in our readers, so this is an attempt to rescue an important part of the life histories while counteracting the potential effects!

So many conventions, then, restrict the translator (Lefevere and Jackson 1982). Readers expect a familiar, conventional narrative from a life history, and anything else creates a barrier for them; yet a conventional narrative in Western terms may grossly distort the original story if it has been told in different narrative conventions, embodying different ideas and images. Readers – and publishers even more – tolerate a limited length of academic book, for cultural and economic reasons, and are used to interview texts in which much of the repetition of words and phrases integral to oral communication has been edited out. Most life histories are presented as 'fake' monologues, put together by the editor from perhaps dozens of interviews to imitate a single narrative. Readers in the West have been encouraged to accept the life history as the direct, unmediated voice of the narrator, as a message from another culture, personal or collective; to trouble them with thoughts of the difficulties of transculturation or the many ways in which the context of an interview determines what is said (Personal Narratives Group 1989; Langness and Frank 1981) may actually deter them from reading it or, ironically, devalue the importance of attempting to listen to insider voices.

The limitations of space forced us to include four, not twenty-six, life histories here, and to choose painfully between two complete life histories and our final four stories, each with half its words edited out. My desire to show the editing process as fully as possible collided again with lack of space, since the more I tried to indicate and summarise our cuts, the less we could include of actual translated life histories! We use endnotes, but how many readers will shuttle back and forth to them? In the end, Janet and I decided to limit endnotes to four key areas: explanation of those local terms which we saw as particularly important; repetitions which we have edited out, losing the emphasis of the repetition; examples of the most difficult passages to reinterpret, with a brief explanation of the nature of the problem; moments when a shift in expression seems to suggest a troubled relationship of the speaker either to the experience itself or to its positioning in language. In other words, points at which she seems to be struggling between competing images of herself and resisting the discourses of the more powerful.

The challenge of translation as transculturation, as set out here, is fraught with so many problems that it may seem overwhelming. Ultimately, however, we are dealing with situations of unequal power where communication – understood here as a thoroughly mediated process – is none the less desirable, even compelling. The women whose 'voices' you are about to hear have their own reasons for choosing to tell their stories, some of which they expressed in terms of the desirability of women getting together and co-operating more, some in terms of possible interventions from outside. There may be other needs and desires at work, about which we can only speculate, such as satisfaction, solace or pleasure in the narrative process of self-construction and analysis as self-explanation to a limited interlocutor perceived as trustworthy. For our group of women researchers, the process raises issues of relationships and representation which we would like to see at the centre of the whole education process and beyond, in the context of interventions by agencies of various kinds who may have the power to meet at least some of the needs identified by the women who have spoken here. This book does not pretend to answer many of the questions it poses but we hope that the questioning itself will be fruitful in its disturbances.

8

CARMELA'S LIFE STORY

Carmela is a widow of 40 who is an ejidataria *and has a small shop in her house. She lives in Plan de Ayala, a very poor settlement in Los Tuxtlas, with her five children.*

What was your childhood like?

Well now, as I was telling you when we were in that place, when people threw out those of us who were poor from San Juan Seco, in the San Andrés municipality, only the pioneers were going to be allowed to stay, only the ones who already had land. And they threw us out with our parents, and there was shooting going on! And we – my brothers and sisters and me, we were really scared and we hid and so . . . Our parents, who knows how they did it? found this forest[1] here and because we were just little, well, we were crying because we were sad that we had to leave, right? And then we cheered up again because we were going to see somewhere new, and they said it was really nice. Well then, our parents and some other people, other families, they got hold of some transport and we came over here, we got as far as Mario Sauza with this kind lady who made a little space for us to squash up together, and almost all we had to eat was plants and fruits from the forest, as we'd gone there with so little . . . and later on they built some houses for us to come over here. My mum said 'Why don't you go over there and get something put up, even just a shelter?'[2] because my mum was just about to have a baby, he ended up being born here that baby, and my mum said 'It'll come any day now and I feel embarrassed, I'm having a baby and I don't want to have it here.'

So then my dad brought us along to help him carry the palm leaf and the sticks and so on, and we made a tiny little house, a teeny weeny one, we couldn't make proper wooden walls, just fabric, and then my mum came to have the baby here, that was in '25 . . . no! In '65! because my brother's 25 now and he's the one who came to be born here. So then when we got here we felt safer but soon we were afraid again because the folk in La Palma didn't want us here, they came along armed with sticks and other weapons, to frighten our parents so that we'd get out of here! But then my dad said 'When those people get here, everyone, women and children, too, we want you all to back us up' – he says –

144

'because we're going to put up a fight; when those people get here, you stand your ground, you say to them, if they're going to kill the menfolk, they'll have to kill everyone! And if they want to take us to jail they'll have to take all of you, too!'

So that's what we did and it lasted a long time and that's how it happened, but then they made peace, our parents got it all sorted out. They even built schools for us and they got it all set up, they put us into school, they asked for teachers, and the teacher who taught us was called Fortino Pales Ronson, the one from Jalapa, he was a patient man, what he put up with from us lot! We had a wooden school house with a roof made of bark[3] and palm leaf and we used to write on top of planks, he put some planks out to use as bench-tables and we would put our jotters on them and the rain used to bucket down! And . . . all we had to eat was tortilla with a little bit of tomato, that's what the teacher ate too, because in those days we couldn't get out to the town. There wasn't any road, just a forest track, and we were frightened of poisonous snakes, because there were lots, no one went to town very often, this was deep forest back then, everything was closed in around here, you couldn't see daylight, it was like it was always night, and it rained every day, our parents planted maize and it wouldn't grow. Well the cobs grew on the corn, but it all sprouted when we went to pick it and it was all stalk, and they used to plant beans, too, but really we only managed to survive because there were little crabs and snails and we ate all that kind of thing, there was *flor del chocho*[4] and we would cut some of that and . . . that's how we kept going, until they found out that dwarf tomatoes would grow well there, being a forest, the tomatoes like the fallen branches, so ever since we were little – my mum brought us up to work – she would say 'We're going out to pick tomatoes', and the eldest would stay behind to cook my father's meals and I . . . because I was the youngest she would take me with her to sell them[5], she would make me carry a big tin of them, about 10 kilos it was, and she would carry another one and then a bucketful on top of her head, and off we'd go on the forest track to Dos Amates, we'd stop off in Dos Amates to catch a bus to Catemaco, we'd go selling door-to-door and sometimes I'd say to my mum 'Mum, I'm tired, my back hurts!' and my mum would say 'I know, dear, but we need to eat, you'll have to put up with it', she would say, 'because you know lots of people suffer and they still don't have anything, but we eat because we work, and when you ask me for money I give you some because you've helped me . . .'

All my childhood it was the same story . . . selling fish, tomato, limes, bread [. . .]

How old were you?

I must have been . . . about . . . I started going out to sell things when I was 7, I did it all through my childhood, I was out selling until I turned 12,[6] round here, over in the Laguneta region . . . you know . . . San Juan Seco, we went back there because we knew people there and we did manage to sell a lot there. My dad would do the farmwork and we'd be out selling to get enough to eat . . . Then my mum said 'Look, we've saved enough to get a proper little plank house', she said, 'find out where they'll cut the wood for you and we'll put it up.'

You see, there weren't any electric saws back then, just that one where you push it backward and forward like this, so my dad sent off for the wood and he built us a proper little house and so, you know, we were happy, right? Because the house was better and then when I turned 13, my dad – my brothers were growing up too, so I helped him out more and I had a bit more time to myself. But at that age, so to speak, you fall in love easily, ay me, sometimes out of desperation, as we say, 'I'm having a hard time but perhaps if I get married things'll change', well, that's a load of rubbish, it gets worse!

Anyway, when I was 13 this boy began to chat me up . . . so, well, I liked him, only I said to him 'OK, but wait until I'm 15', I said, 'and then I'll marry you, and he did, he kept an eye on me, but then when I turned 15 I said to him, 'OK then, go and speak to my parents', and he said 'I can't, I'm too poor, I don't have any money', he said, 'how can I get married? Weddings cost a lot of money.'

So then I said to him 'What do you mean, don't you love me any more?'

So he said 'Yes, of course I do', he said, 'but why don't you just come and live with me?'

So . . . well I was in love with him so I said, 'OK then, I'll do it, let's see if this changes my life.'

But you know, we were already a bit happier because my dad, em . . . the maize was growing because the forest was being cleared and we were more settled . . . anyway, I fell in love and off I went with him. We got married later and we carried on working, we lived with my mother-in-law for a year, but then I said to him 'Ask your dad to put up a house for us next door and let's try it on our own, to see if we can make a go of it', I said, 'because, you know what, they're saying in the bank that they're giving loans, so I'd like us to plant some crops of our own, right?' I said, 'because when we plant crops with your dad, he only gives us something if he feels like it, and', I said to him, 'I married you because I wanted some say in things, too.' And so he says, 'You're right, you know!'

So he speaks to his dad and his dad was furious because he didn't want us to go off on our own and *he* says 'I don't have enough money to build you a house and even if I could give you the house I've nothing for you to put in it.'

So he [her husband] says, 'Don't bother with the gear,[7] just give us the house.'

So he makes us this little shack, 'Room for two cats here', he says.

So then I said to my mum 'What am I going to do mum, here I am with a house but nothing to cook with.' And mum said 'Don't you worry, I'll give you a little something', she said, 'I'm going to give you each your cup and plate because everything comes from small seeds and then little by little . . . ' she said.

So we set to work, him and me, planting the maize . . . he cleared the land

and I did the other jobs, sowing, weeding. And the first year things went really well, you know, we paid off the bank and built a much, much better house . . . then we had our first child and, well, things got a bit harder for us, because when I had the baby, it didn't go right, I felt ill – anyway, then, after the second, we had a little cow by then, we'd bought it with the harvest money, anyway I got a lung infection,[8] so my dad took me to Jalapa and I was in hospital there. So there I was having this treatment and I'd had to leave my little boy behind, only 9 months old, he was still breast-feeding. I told the doctor, so then they tell me that when I go home they'll be giving me my treatment in the health centre in Catemaco, so I come home and I find my little boy ill, really ill, with lots of sickness and diarrhoea, and they wouldn't let me hold him, because they said I was too ill and I was crying my eyes out, because I was thinking to myself 'Is he going to die?' And you know, he did, my little boy died and my husband sold the cow to pay for my treatment, and the maize harvest, and I used to say 'Well, you know, I've lost one child but at least the other one's all right, because if I had lost them both' – that's what I used to say – 'then what would I do?'

. . . so . . . life went on, I started to get better, I got back on my feet, more or less, and I could go and fetch my medicine [. . .] I also said to him 'We'll get out of this mess!' And we did, we went on working hard and we got back on our feet, we had a good harvest and we bought another little cow and . . . uh, there we were, I had some more children and I finally got over my illness and things were just looking up for us when *he* died [sighs]. That was so hard, it really was, because I was feeling pretty good, you see, with the other children I'd had by then and then he said 'I think I'd better send you to be operated on', he would say, 'I'm going to take you for the operation, because it's not good to have so many kids.'

How many children did you have?

I had seven, so he says, 'I'd better get you seen to so that you and I can both get a bit of a break.'

So I said to him, 'OK, if that's what you want, what matters is for us to keep working.'

And that's what happened, he decided to have me operated and he took me to Catemaco, I didn't want to do it but he said 'You must do it', he said, 'you've got your hands full with them already and then you come and help me out and I don't want you running off after them, and so they suffer and so do you', he says, 'because you have to leave something for them to eat or rush home to prepare meals.'

I used to take my kids out to the *milpa* with me, I'd find a shady spot for them and put up their hammock, because I've always liked to be doing something, to help make ends meet, so then when he made his mind up, well, then I said 'That's that, then.'

So then he took me away and they operated on me and afterwards I said to him 'Well that's it, I've had the operation, we've had our last child, there'll be no more babies now, we can work hard and the kids, the boys, can help us', I said, and we had two cows by then, fully grown, so I felt more secure, I said that I wasn't going to face any more hardship, I was sure of it, when suddenly he died, he fell from a house, he was building a house, when I went to get him he was still moving, and we rushed him off to San Andrés, to the health centre, and from there they sent us off to Jalapa, and then I said to myself 'I don't care if it takes everything I've got, I don't want him to die', so we went to Veracruz, the port, and I said to myself 'What am I going to do? How am I going to do it?' And I'm feeling that if he dies I'll never cope with the children and yet, when he did die . . . well, I just coped with it, that was about five years ago, ever since then, for those five years, I've done it all alone, the kids, working like crazy, life's been pretty hard for me. When he died I was left with a huge debt because I'd had to borrow 400,000 pesos to pay for everything, because he had to have an operation.[9] I fell ill when he died, I had a nervous breakdown. I was in bed, I remember my doctor said I was ill in bed for about a week, they tell me I would talk but I was, you know, delirious, and maybe that's what caused the problem, because they gave me drugs to make me better, and then I was feeling really awful, even though I'd been on my own for six months, I felt terrible [. . .] I was in the clear. Well then there was a robbery round here, they stole twelve cows from this man, and I was afraid someone would come and steal mine, so I said to my dad 'You know, I think I'd be better off buying a little maize grinder and working for the local farmers, because they've just robbed this man and I'm an even likelier target because I'm a woman living on her own, and you know how much we sacrificed to get them, we've gone without food to do it, and I'm not going to have them stolen off me after all that', and he says 'You know what, that's not a bad idea', he says [. . .] I thank God that since I was left on my own, I've never gone without food because I've always been a fighter, I grind everybody's maize and then I go off to Catemaco to find the bargains, and that's what I bring back to spin out the money. So that's what I've done for work, and I may be tired but I'm happy, you know? Because my children are growing up now, the oldest is 16 now, the next one's 12 and they're not a bad lot, you know, they're not bad, I say to the oldest 'You're a big lad now, you're quite a help to me now.'

Because although it's harder, because the bigger they get, the more clothes I have to wash, and they ask you for more, because one lad sees what another lad has and he wants one too . . . well if I've got the money I'll get it and if I don't I just say no, because it costs a lot to run the household, right, and I've had problems with my nerves, you see, I'll be fine some of the time and then suddenly I get these headaches, and I've seen a lot of doctors about these headaches, because I've been like this since my husband died, you know, and all the doctors say that it's a nervous thing, and they still haven't cured me! And what I say is, well if it's a nervous thing how come they can't cure me, right? I mean there must be some medicine for it, right? And then the others, they say

148

'Look, in our opinion, you're not going to get better until you get married again, you need a husband to help you to think straight, to help with everything.'

But what I say to them is 'What do I want a husband for?' I say, 'maybe I *need* one, but when I start to think it over, well, I mean when the father hits the children, well, he's their father, right? But to have another man come along and raise his hand to my children, well I couldn't put up with that.'

And then I start thinking . . . 'Yes I do need a man, but my children are growing up and I could never, for example, I'd never let a stepfather lay a finger on my children, when you've looked after them since they were just a little bundle, and all you struggle through to bring them up, and then when finally they're grown up, to have another man come along and take advantage, you see what I mean?' So then I think, 'Well, I couldn't put up with that and I would have a lot of problems' . . . and that's how it's been, that's why I've never remarried, for my children's sake. And then there are some men who only want a woman because she's on her own, they see me making my own living, wanting for nothing, they see that I've got work, but if I got married I'd want my husband to keep me, I don't mind helping, but just up to a point, because at the moment I have to work far too much. I mean, on the rare occasions that I go out, to Catemaco say, I won't eat myself so that I can bring something back for them, because I'll say to myself, if I have something to eat here, they'll go without, so it's better if we can eat something together later. So what I say is, what do I want someone for if he's going to expect me to keep him, right? I mean he ought to be keeping me, right? Sometimes I just wish it would be God's will to send me someone, right? Someone who would more or less take the place of my husband, not quite the same, because I suffered a lot of poverty with my husband but he never hit me once, not even when I disobeyed him, I suffered poverty but he never hit me, none of that, in that respect my life really was happy, so what I say is 'When am I going to find another man like him? Never', so that's why I'm still on my own.

And your children are all boys?

Four boys and a girl.[10] So I tell the boys that they're men, they must learn to work [. . .] They get desperate, because I've seen lots of uncared for children who go around stealing from others because they've been neglected, right? Not me, not me, I say to mine, you should never take something you take a fancy to, never! The day you need something you want, you come to me and if I've got the money I'll get it for you and if not I won't, but you must never steal, because that's the saddest thing, if I've got the money I eat, if not, I don't, so in that respect I'm happy because they do what I say [. . .][11]

Yesterday, thank God, I sold a lot, I went to Catemaco and, thank God, I sold everything I brought back, I brought soft drinks, quite a lot, and they were selling like hot cakes, I brought shrimps and mangoes and everything and I said to him 'Look, son, today I made 30,000 pesos [£6 or US$10], it's not every day that I make that much, some days I only make 5,000, and what's 5,000 pesos?'

But that's how it goes, right? In business, you win some you lose some, you don't give up, you have to say 'Today I made a loss, but tomorrow I'll make a profit, because it's never the same from one day to the next' [. . .]

And how long have you had the shop?

Since my husband died, that's when I set it up [. . .]¹²

And would you have liked to have more children, more daughters?

Well, you know, I would have liked more daughters, they were going to operate on me when I just had the four [children], but I said no, not until I have a girl, and finally a girl was born, so then he said 'That's it, otherwise, if the next one's a boy, then when will you have it [the operation]?'

Now I wish he hadn't had me operated, you know, but what can you do? Can't do anything about it now.

So you would have liked to have another girl?

Yes, another girl to help me around the house.

How old is your daughter now?

My daughter? She's 9 now.

And does she help you around the house?

Yes, yes she does, when she's not at school. I give the house a good sweeping, I give the cloth a good wash and I give her her bucket and I tell her to get on with the cleaning and she does, she does the washing up too, and her own washing; well she does if I tell her to but not always. Do you know she tells me she wishes she was grown up so that she could help me, and she says 'When will it be? When will I grow up?' And I say to her 'You just wait, one day you'll grow up.' And she says to me 'Even though I eat I don't grow', she says, 'I'm going to eat lots and lots, so that I grow up so I can help you.'

And how are they getting on at school?

Well they've done all right, not bad, because sometimes kids don't learn because they don't try, because you know, the 16-year-old, when I was left on my own [widowed] I took him out of school, because who was there to help me? Who would water the horse? So I took him out, but you know what? He knows more than this lot.

So he's not thinking of going to secondary school?

Well, you know, I'd like him to but I can't afford it, I've got one son finishing primary school this month and he does want to go on to secondary school, but I can't afford it.

But there's a telesecondary school in Dos Amates, isn't there?

Yes, there is, but like I said I can't afford it. What can I do? So I said to him 'I'm going to buy you your machete and your hoe so that you can go and work in the fields, because what alternative is there? There's no money to pay for more studying', I told him – 'at first perhaps, OK', because it gets more expensive the further up the school they go, so I tell him 'If I send you to secondary school I could manage at first, but as they get further and they get more qualifications, they need more expensive things, and that way, I don't make an investment, because you invest in them thinking they're going to get a profession and if I'm not going to be able to pay for the whole course, well then it's better not to give you anything so that you don't waste your time.' He does want to go on, but I can't afford it [. . .] I have a hernia and the doctor says I need an operation, but I'm not having it done because it costs a pile of money.

Did you get it from lifting things?

Yes, from lifting things, I went to see a doctor the other day and he told me 'Look, I'll help you out, if you can get hold of 500,000 pesos, the assistants cost 200,000 pesos and I'll pay that bit', he said, but all I can say is, 'How on earth could I do it? Only if I starve myself' [. . .]

I would like to go and work somewhere else, I've told my parents I'd like to leave my kids with them while I go off to work, to get to know some new people as well, and I told them, I'll work for everyone, I'll go wherever they'll pay me a good wage, because there are some jobs which pay you well without half killing yourself like I do here . . . so I said to them 'I'll go and work for everyone, I'll send you part of my wage', but they didn't want me to go and work in another town and get to know other places. I said to my dad, not recently, when the little boy, the last one, was younger, I said to them 'Why not?' And they said, 'Because any woman who abandons her children and goes gallivanting off to see what turns up, is a loose woman.' And I said to them 'Look, any woman who wants to can go astray, but the woman who doesn't want to, won't, because any woman who wants to will do it, even in her own home', I said, 'and the woman who wants to look after herself, will, even if she's out and about on her own.'

Yes, you can make yourself look nice not to be attractive to a man but to feel good about yourself, right? Exactly! I – they say – I've heard them say on the radio, 'Make something of yourself!' Now and again I've listened in to that programme *Nuestro hogar* [*Our Home*], I like it because they give you advice, they tell you not to let yourself be bossed around by anyone, go on and assert yourself! You have to do it yourself, because who's going to say that I'm all right? It has to be me. So it's true then, unless I can make something of myself, unless I take care of myself, who will? That's what I think, but not my father, not him [. . .]

Which religion are you?

Adventists, yes, as I was saying, it's very important for me, because my son doesn't drink, he doesn't smoke, for me, my son helps me, because let me tell you, there are lots of young lads around here from the Catholic Church and on Sundays I see them going around well and truly drunk, really legless, going around yelling, with their ghetto blasters blaring, there are youngsters shooting off their guns, so I tell him, my son, the oldest lad, 'Stay away from that lot, because they're drunk, right? May God prevent it, but a shot in the air, you're just passing by and you get it, right?'

Well, he doesn't do it, he doesn't wander off at night, he'll go out and be back at 9 or 10 p.m., or sometimes, I've got a sister over the way who has a TV and sometimes I'll ask him 'Where have you been?'

'Over there watching the telly', he'll say, so my son does go out but I'm not left here worrying because I know he doesn't drink, because a drunk, well, he's a danger.

What do you suggest, something to improve the situation of women in this ejido*? Something you would like? A school of some kind?*

Well, you know, it would be nice to be able to study nursing, and to learn dressmaking properly, because that's the kind of work you can do at home and it would also be nice to learn to be a beautician, you see, me now, growing up very poor, right? Suffering a lot, I've always liked to have a go at everything, in my house, if I hear a rumour that over in Perla de San Martin[13] they're going to have a dance, I'll go to the authorities to ask permission to sell something.

'What are you going to sell?' they ask me, so I tell them 'What do you want me to sell?'

'Well, in that case, you can sell food', they tell me, 'but no beer or soft drinks, because the works committee is selling those.'

So that's what I do, off I go, I get hold of some transport, I take my stove, my gas canister, my pots and pans, I take everything and I look for some girls to help me out because, you know, they're best at selling, so I tell them, look, when people start arriving, I tell them 'You get ready to sell them whatever they want', oranges, mango, sweets, soup, and that's how I make my money, I get permission and I make my money, of course I'm up all night selling and then early in the morning I grab my things and my transport's there waiting to take me home.

It's been said that for things to improve the other women need to go out to work, too, but who can help them deal with their man? Lots of men beat them, spend their money on drink, what can the women do? Well I'll tell you something, an intelligent woman, one who can think, well she can make her mind up, can't she? I suppose, because a woman who doesn't like working, even if there were enough work available, doesn't work because she doesn't want to, I mean I've seen some women sitting, just watching the world go by, and coming from somewhere or other I think how relaxed they look, how happy they are, sleeping in their hammocks in the middle of the day, and I think, well, they don't need to work, I think the only one who needs to is me, but, I mean, there are some women, their stomachs may be crying out for food, and there they are, resting in the shade, and whether they have food or they don't, it's all the same to them.

And if they worked, do you think their situation would improve?

Well, yes, I do [. . .]

What do you suggest for the women of the ejido, *to improve their situation?*

Well, for us to get together to discuss things and to ask for a supermarket, a CONASUPO,[14] because if we're to get together the *ejido* needs to be big and to get organised and ask for a plot of land for the women, but the *ejido*'s so small just now it hardly exists and we don't achieve anything, but we could get organised and ask for the things we need.

Life story collected and transcribed by Ursula Arrevillaga and Janet Townsend.

NOTES

1 In these areas, *montaña* means either mountain or forest, land not being used.
2 She uses *manteado* for 'shelter', possibly a piece of cloth or canvas.
3 Roof made of *jonote*, a tree with a very thick, strong bark.
4 A local fungus.
5 To the *rancherías* or scattered clusters of houses, hamlets.
6 It would no longer be respectable to be out selling once she had passed puberty.
7 'Don't bother with the *trastes*' – plates, cutlery, pans; the means to cook and eat.
8 Tuberculosis, in fact.
9 Her father helped her.
10 Two men died.
11 She talks about her son, who wants to marry.
12 She talks about how she used to buy *barbasco*, the wild yam from which some contraceptives are made.
13 Nearby village.
14 Government shop for cheap food.

9

ELENA'S LIFE STORY

Elena, aged 44 when she told us her story in El Arroyo, was born not far away and moved there after her marriage. She completed three years of primary school and can read, write and count a little.

When you were small, what did you do to help your parents?

Well, I used to help my mother with the cooking, with household chores like cleaning, washing clothes, preparing food, grinding [corn, etc.]. Because I was the only girl they used to spoil me a bit, but my mother always got me to help her grind the corn in the kitchen. And when I was about 7 years old, yes, about 7, she started me off cleaning, putting out the rubbish, washing dishes, feeding the animals and so on.

You said that you were the only daughter?

Yes, my mother had about sixteen children but lots of them died, ten of them died. So out of all of them only six of us were left, my five brothers and me [. . .]

When you were about 8, where did you live?

With my parents, over in Zaragoza, it was after I got married that I came here with my husband. I got married when I was about 14, when my first daughter was born I was just turning 15. That is, I gave birth in July and turned 15 in December. Yes . . . I was just a kid! [. . .]

How did your parents treat you when you were little?

Well, my father treated me really well, with lots of affection, lots of love, lots. My mother too, only you know, well you know what mothers are like, we're very disobedient and, ay me, we want to smack them.[1] Yes, my mother punished me now and again but they always loved me a lot, since I was the only girl, yes, they really did [. . .]

What did you most like about your life as a little girl?

As a little girl . . . playing! To be with my friends, to do, well like during Holy Week as we call it, I loved to play with my friends. I would invite them over to my house, we would make snacks. I had some cousins, yes cousins and other boys, and I would say to them too . . . 'Come on, let's go and play at cooking!' and my mother would get me to make little things out of clay, little pots, a frying pan, that kind of thing, to play with, and they would make little houses and ovens, mini ovens, and I would make my kitchen. And we would play with my girlfriends. I loved to play with my friends!

What other games did you play?

Well, just that really, and the ball game I told you about [baseball]. I really enjoyed school and I got on very well with my friends. I hardly ever quarrelled with them because my mother didn't like me coming and telling tales on my friends. She was always saying 'Don't you be coming to me telling tales about your friends' . . . And sometimes I would come to her and say, 'Mummy, do you know what my friends did?' 'No! don't come to me with your tales, because I'll give you what for!' . . . She wouldn't let me tell her anything.[2]

Did your mother ever punish you?

Yes, she would punish me sometimes, if she sent me on errands and I couldn't be bothered, or she sent me on an errand and I bumped into a friend and we'd start chatting . . . then it would be 'What shall we play? What shall we do?' When I was bigger, well, then we'd talk about boyfriends. But she always punished me because I'd hang around, so I didn't get back quickly enough from my errands.

What do you remember of your childhood that you didn't like?

Let me see . . . what didn't I like as a child? Well so far as playing is concerned I liked everything . . . What I didn't like was seeing my friends quarrelling, gossiping about one another. No, I didn't like that and I still don't.

Anything about your childhood that you don't like, any nasty memories?

I haven't any . . . really no, I'm sorry. Nothing I can remember.

You say that your mother used to tell you off or punish you, how were you punished?

With a rope, a vine they used for tying up firewood, Ay me! They always punished me with that!

Did it hurt?

Yes . . . ha! ha! sometimes it hurt!

Were you afraid of your mother?

Ah! Was I afraid! . . . I was really frightened of her, even when I was older, until I got married. I was afraid of her because she used to punish me. You see, she didn't like it, even when I was older and unmarried, when I went out with boys, she didn't like it at all. You see, sometimes when you're with boys like that, you all start playing these games . . . and she didn't like me playing with boys, she always told me off [. . .]

What did you have to do when you were about 12 or 13?

Well it was always housework, cooking, because my mother . . . they were the kind of people who thought it was important that you learned to do things, and although she was at home, she used to do other things, making clothes, sewing, and she would leave me in the kitchen and she would just say to me 'You're going to do this and that.' She would tell me to make the *tortillas*, to grind the maize, wash the dishes, clean the kitchen. And lots of times she would come into the kitchen and find me there with the kitchen all dirty and boy! would she tell me off. Sometimes she would punish me. She was always concerned that I should learn something. One thing though, there was one thing, she always gave me mending to do, and embroidery, just a little bit but I did learn how to embroider a bit, to sew, to do a few things.

At that age were you already going out with boys?

Yes. I began going out with boys when I was 11.

How old were you when you started to menstruate, to have your period?

Eleven, yes I was 11 when I had my first period.

Did you know about periods?

No . . . I didn't know! Ha, ha, ha. I got quite a fright! *Híjole!*[3] When my first period came my mother wasn't there, she was with one of my uncles who was very ill. They took him somewhere to a healer[4] and I stayed with my grandmother. [Coughs.] Well, that day I felt awful, my head ached, my legs ached and well, no, I didn't know a thing until the evening when my grandma said, 'Right, Elena – when are you going to bathe? It's getting late my dear, off you go.' So I went off for the water, got the cubicle ready and there I am just beginning to ladle water over myself, but suddenly I see that I'm . . . *Híjole*! I really got a shock! I mean, I don't know what I felt. When I saw THAT I said: 'What is it, what's wrong with me?' Ay, dear God, I didn't know what to do! [. . .]

And what advice did your grandmother give you?

Well, she advised me that when a woman reaches this, this thing . . . well you have to be careful not to go with boys, not to let a boy do you wrong, sort of saying if a boy is courting me, not to go and let him take me off somewhere on my own, to be with him.[5] So she started to tell me all this. And about shame.[6] That this is a woman's shame, so when you're like this, you've really got to take care of yourself, dress warmly and put something on so that you don't stain yourself[7] because if you go out stained it's very shameful. She began to tell me all this. Yes, all about it.

So your grandma talked to your mother about it and then your mother talked to you?

Yes, then she told me too, she really warned me about being very, very careful, with boys, with my girlfriends, because she even said, 'There are always friends who say they're your friends, but then someone invites you out somewhere and they leave you alone with their [male] friend or something, they can lead you on into all kinds of things.' So it was really drummed into me.

And then you began going out with boys?

Then I began seeing boys and . . . no, they didn't manage to lead me on because what my mother, what they told me stayed with me and . . . [. . .]

What is the age difference between you and your husband?

Well, he's 40 something . . . [Asks her daughter, 'You don't know how old your father is? Ask la Lencha'.] . . . Ha, ha, ha. Yes he's about 40 . . . no, 54 I think. Yes . . . he was older than me, yes, because when we got married he was . . . he's 55 now.

And how was it when you got married? How did you get on?

Ah, very well. Very well, but you see I . . . after he spoke to my mother and my father and we were going to get married and everything, I felt that I was beginning to fall in love with him and then we got on really well. When we were going out together we never used to fight like other couples, out of jealousy, because he had other girlfriends. Well, I felt it didn't really bother me that he had other girlfriends and . . . well, then we got married, but then on my wedding day I felt terrible, I cried a lot. *Híjole*! I cried and cried. There I was on my wedding day and I've got a photo of when we got married. There I was crying, crying my eyes out! [. . .]

Yes, after we were married we came over here, we spent about a week at home. And then we came away, and about a fortnight, or a month later, he took me back where my father lived and I felt that I didn't want to come back here, because I felt, and I saw, the place – it was horrible! So dark at night, surrounded by forest, it was all around, *híjole*, it was awful! At night, when darkness was falling, ay! I felt I was going to die, it was terrible . . . really awful . . . so sometimes he would take me to my parents and we would spend a few days there.

But he treated you well?

Yes, he treated me very well. Yes, I can't complain, he never . . . well, whether or not he treated me badly, on the other hand, he would give me advice about things I didn't know how to do. He would show me, he would tell me how to do them, or I would ask him, 'Gilberto, how do I do this? Tell me.' Or he would send me to see to something, some food or something. And I didn't know how to do it – well, of course he had brought himself up really, I don't mean as an orphan, but when his mother died he was only about . . . 15, 15 or 16, when his mother died. But he had some younger sisters and you see . . . there he was all alone with just his sisters and well he knew how to cook a bit and that kind of thing. Whenever I felt that I couldn't do something, I would say to him, 'How do I do this?' 'Ah, well, you do it like this.'

He treated me very well, he was very affectionate, even if, as the years went by – and nowadays, too – we fought a bit.

And what was your first child?

A girl. Well, no, what happened is I became pregnant and at three months I was damaged. I had a miscarriage. I didn't know, I didn't know that what was happening was that I was pregnant, because my mother had never told me about it. But when I was, when you're pregnant, then I noticed that my period stopped . . . well, then, at three months when . . . I felt ill again, but I didn't know what was going on and he, well he didn't have a mother, his father was alive but he worked on a farm, he didn't come to the house. So he . . . didn't have anyone to tell, to ask either. Well then . . . some other women asked me if I was pregnant and I said, 'Who knows?' . . . I didn't know until afterwards a woman said to me, 'You see, dear', she says, 'when you're pregnant your period stops.' She began to explain it to me. But I'd been feeling pretty ill for about a month then, when they took me away from here, first to a place they call La America, where he was working then, and from there they had to take me by aeroplane to Zapata, I was really ill, seriously ill. So who knows if my first child was a girl or a boy? Later I fell pregnant with my daughter, so my eldest was a girl [. . .]

And did your husband tell you what he wanted? Girls or boys?

He didn't mind which it was, it was all the same to him, girl or boy, he was very happy whenever a girl was born, or a boy, they were almost all girls and there were . . . nine, nine or ten girls [asks her daughter, 'How many of you are there now?']. Six, but counting the married ones . . . nine. And one girl who died: ten. And two of my boys died and two are alive, so, all in all, I had fourteen children, ten girls and four boys. After the fifth girl he started on with wanting a boy, a boy and . . . no boys! [Laughs.] Until finally the last one we had before her, that was when we had a boy. And we really loved him! He was very happy, very contented that I'd had this boy, and everyone, the girls too, everyone was pleased. I felt very happy but then I had the bad fortune to lose him, our little boy died [. . .] So my little boy died and then I had Jasmine. Another girl!! So I went to . . . I never had any of my babies with doctors, always with the midwives, everything! But after this girl I went to the clinic. Because he thought I shouldn't have any more children, girls or boys, so I went to have the operation.

How long ago was that?

The same as my daughter's age, ten years.

And did you decide that together or did he decide?

Together, we both decided [. . .] So . . . I said [. . .] 'Well, let me have the children that God gives me', I said to him. But then he decided, 'No, sweet-heart', he says, 'don't you see that there are enough kids already, too many already, let's have you operated on'. And off I went to have the baby and there . . . they did the operation on me [. . .]

And after you were married and living in your own home, what else did you do apart from the housework, just the housework?

Well, apart from the housework, I . . . I've always liked helping my husband, working, making things to sell, so I was always making sweets, *empanadas, tamales,*[8] things to sell, bits of clothing, because I made clothes with my sewing machine, I made girls' dresses and knickers, women's dresses to sell. I'd take the stuff off to sell, I've always enjoyed that, this business thing. I'm always selling clothing, always selling something! Now since we moved over here, now that my daughters are growing up, we've sent them all to study, not a great deal, but we've taken the trouble to see that they learn something, that they do their

159

studies and that. And then, well, I've devoted myself to helping my husband with work, even just selling bits and pieces like this, I've sold Tupperware, Avon [make-up], that sort of thing, food when the teachers come to do their classes – I don't know why, but they've always liked to come here to the house [. . .]

And you didn't work in the fields?

Oh no! [Laughs.] Not in the fields!

Do you have a plot of land?

Yes, my husband is a member of the *ejido*.

Do you ever go out to the fields?

Sometimes, when he's got to dip the cattle, when he's clearing the weeds, sometimes I'll go with him. Just for an outing, not to work, no way! You see, since my father didn't get me used to the fields – my father worked on the land, he grew fruit, he had his *milpa* and so on, but he didn't really like me going to the forest at all. So my father, well there are men who don't like to be lugging their womenfolk along to the forest.

And do you like working in the fields?

Weeelll . . . Who knows, I don't really think so – what I do like, sometimes when we go out to the fields I'll start clearing a bit of woodland, collecting firewood, that kind of thing but not like some women who like hard outdoor work, ploughing, weeding. Of course we always help him when he's preparing the maize plots, clearing the land, harvesting the maize, picking the fruit and all that [. . .]

What kind of advice do you give your daughters?

Weeelll, all that business about how they must make an effort to learn, to study, even if it's just a short course, so that when the day comes they've got something to keep them afloat, for instance if they married someone who's not very well off, if they don't have . . . well, then they would both be poor. So if they're going to get by, well, then she's got to learn something, because they'll need it later on [. . .]

And how do you get on with your husband now?

Weell at the moment, I think it's a bit up and down . . . some days we're quarrelling, some days . . . we're happy, like sometimes I get in a bad mood and then suddenly, we're bickering! But when they, my daughters, take me up on it, him as much as me, they say to us, 'No, mum, you two shouldn't fight, it's horrible. Versabe told us that it looks awful.' And then they tell me that they're upset, that they feel bad when we fight, seeing us angry, well, then if he's not talking to me sometimes I won't talk to him either but then sometimes I feel: 'My poor daughters, they're all upset.' And then when we start talking again, or I talk to him, or he talks to me . . . then we make it up and they're so happy to see us in a good mood again they even cheer us on, and then I think 'Well, OK then' and that's that.

And why do you fight nowadays? Does he drink a lot?

No, he's not a drinker. Well now and again, when there's a fiesta, or a party, or something like that. Or sometimes when he goes into town to buy, then he'll have a drink, but it's not as if he comes home drunk, falling around, no . . . well, sometimes, over things like money, sometimes we're a bit short, and we get . . . well, bad-tempered, I think that, you know, because I, well, we women, me, I . . . [laughing] when I'm short of money I get angry, it makes me furious, if I don't have enough money to buy the things we need, and then sometimes he . . . gets angry with me if he hasn't any money and I ask him for some, we've run out of such and such and he says to me 'Well, what do you expect me to do about it woman, if I haven't got any money?' And so . . . we start quarrelling about it.

And has he ever tried to hit you?

Well, not now because the children are grown up, but in the early days, boy did we fight! We would come to blows over the children, because, you see, he never approved of hitting the kids, even . . . shouting at them, it's not his nature, you see, he's very quiet, very calm, and that's how he talks to the kids. What used to really annoy him, in the early days, when the kids were little was if I hit them . . . because we mothers are always dishing it out, and it made him furious, and then if we started quarrelling he used to hit me and the two of us would start laying into one another!⁹

Did he hit you hard?

Yes, sometimes he hit me really hard, because he rode a horse, so he used these whips, that's what he would hit me with. I talk to my daughters about it, I tell them how he used to beat me. So that's what we used to fight about, because he didn't like me hitting the kids, or . . . whatever, but not other things, his food, his clothes, the way I looked after him, he never complained about that.

Did the two of you have a hard time?

Yes, because I was having all my babies and the kids were all little, we had a hard time of it; we had a lot of heartache, a lot of problems! [. . .] Yes we were really badly off. No, I must say, now thank God, even though it's a constant struggle, hard work, with some rough patches, still we've given life to our children, but we certainly went through a lot bringing them up, *hijole!* When we moved here we lived in . . . we had our little house made of palm leaves, a tiny palm-leaf house, half of it with no walls, and cooking on a fire on the floor . . . it was pretty basic, we had it pretty hard, my first kids had a hard time of it.

Where did you have the kids, did someone help you give birth?

Here, at home, while my mother was still alive I would go home to her to have them, she would take care of me and keep an eye on me, but then with the others, once my mother died, that was it. And I began to feel really lonely, everything that happened to me seemed much worse, and I felt really unhappy [. . .]

And you found it very hard because you were very poor, and how did you find your ejido?

Well when we got here we found the *ejido*, sort of like it was coming round again, because the *ejido* had fallen apart many years ago. Because, you see, the first settlers on the *ejido* were my relatives, my uncles, his [husband's] grandparents, who were relatives of mine, they were among the first settlers, and they managed to get it going, it even had a name, they called it The Little Village. When I was a girl I used to hear The Village this, The Village that. Then later the parents of the Hernandez family came along. My uncle put them up in his own home, they got together and they went on fighting their claim to the *ejido*, but as time went by well the men grew old. You see, when they arrived they were all unmarried, I think, that's what my father told us. Unmarried men who began fighting for their claim to the *ejido*, on and on. Because the first settlers went, they left, or rather they were thrown out. Because they cheated them. So, then they were left and by the time we came along the *ejido* was turning around, because all these people were coming to settle the *ejido*, and they began to seek their loans and all that.

Now my husband had a right to a plot because they gave them to the sons of *ejidatarios*. So his uncles sent for him and he said, well, where we were we didn't have a plot or anything, we were living with my parents [. . .] Away we came over here and when we got here the whole place was still covered in forest. So he began to build his house, a tiny little palm house and half of it without walls, and the doors of the rooms, I mean, I had my three girls by then, all pretty grown up, the door was just boarded with sticks, and, eh, . . . so on. He had to travel to work a long way off, to the ranches (see p. 68) to look for work, and it happened that bit by bit they got the bank to give them a loan and everything. They worked very hard, there was a time where things went very well. So well, that when they paid off the loan they made . . . I can't remember exactly, they made about 150,000 pesos that time. But he thought, he and I thought, 'We're not going to go and waste it, we're going to put up a decent house with this.'

You see in those days everything was a lot cheaper, so off we went, me and him, and he bought the sheeting [for the roof]. After that, we would buy up the materials year by year and finally. . . we built the house! We were lucky because I have a son-in-law who's a builder, he was here with l.is, with my daughter, so we got him to build it for us, about 10,000 pesos he charged, I think, but in those days, it was . . . about fifteen years ago?

You've been telling us how hard you worked and we can see that you really have your hands full: selling things, preparing food for the teachers. Do you never go out, enjoy yourself?

Well, if you mean do I go out just for the sake of it, well, no, actually, I don't! Or if you mean do I say 'Well I think I'll just pop into town', or 'well I'm off out for a bit', just for the sake of getting out, no, I don't [. . .]

And how do you enjoy yourself here? Do you chat to the other women, do you have friends?

Oh yes! . . . I chat to the other women. It's just that sometimes, well you saw yesterday, sometimes we argue and sometimes there are women who, I don't know whether they're jealous or it's in their nature or what, there are some women who don't want to be friends with me. They're always gossiping about me.[10] They always treat my daughters like dirt, they spread nasty rumours about them, they're up to this, they're up to that. Well, that really upsets me. What I say is 'Well I thank God that my daughters, at least the unmarried ones who are here at home with me, well they've never given me cause to say "Oh no, my daughter's bringing a man home", or, "She's got into trouble"' [. . .][11]

And now with the UAIM being set up, is it a boost to the family income, how does the work suit you? Is it a help?

Well, you know, this business of the UAIM, the problem was that that woman didn't take us into account when she got the loan, even though I was a member of the executive, I was secretary, and Doña Anastasia – do you know her? – she was treasurer. And the president – when we were putting in loan requests and going to the bank and so on, because we did go with her to ask for the loans, she would think of us and call for us and off we would go. We even made two trips to Tuxtla and . . . but after that, the day the people from the bank came and told us they were going to give us the loan and sign an agreement with us, she said, 'Tomorrow we're going to Palenque – you're going, Tacha's going, Brigida and everyone.' And, you see . . . well, 'OK', I said, 'I can't say for certain yet because I have to check with Gilberto, but I think I can go, I'll ask him for some money for the fare', I told them, 'let me know when we're going to set off.'

And the next day, I waited and waited for them to get in touch and, do you think they did? Anyway, days went by and nothing happened. Then suddenly, about a month later . . .

'Don't you know Carmen [the president] already got the money, don't you have the cattle yet?'

'Well, no', I said, 'they didn't let me know.'

And then, well . . . more time went by and about two months after we heard that we were getting the loan, ay me! a girl comes to me

'Doña Elena, Doña Carmela says to come and collect the cattle . . .'

'What cattle?' I said to her.

'Well, the cattle they bought with the loan money.'

And I said to her, 'What, you mean she's already bought them?'

'Yes', she says.

'Who went to buy them?'

'My father and Don Gilberto.'

So I said to her, 'Well no one can come just now, there's no one here.' My husband wasn't around, or any of my sons. I said 'I've no one to send, how can I?'

So, anyway, they fetched the cattle, they took them over there, and one of the heifers died on them because it got too much sun and they just fixed it up and sold it. We didn't know anything about it. Well, later the rest of us said, 'I bet she'll get in touch this evening to fill us in', but we didn't hear a thing. It must have been about a week later that she sent for us, so I said to one of my daughters 'Let's go to the meeting, because my *comadre* [pretended relative] Carmela has sent for us and I want to hear what she has to say.' So, off we went, well, there was only me, my daughter, because she's a member of the UAIM, and Doña Brigida, and the other one, Doña Tacha, and one other woman. So I said to her 'This meeting must have been set up to give us our share', and another woman says to me 'What do you mean, give? We're here to make a contribution, goodness knows how it's going to affect us, we have to raise 500,000 pesos which we've just lent out, from the contract we had, we want to renew the contract and we've lent all the money. We lent to . . . who was it? 200,000 pesos we've lent to someone or other and 300,000 to someone else.'

So I said to her 'You don't mean the money's already gone?'

'It sure has, we brought the contract home the other day and we've already bought the cattle.'

So, I said to her 'Well I don't know what's going on, because we don't know anything about it, we were never informed, they didn't say anything to us.'

'Really', she says, 'well, as soon as the money came through they went to buy the cattle.'

So I said to her 'Well, you know what, we didn't know!'

'Ah, well', she says, 'We've all got to co-operate now because we owe 500,000 pesos.'

So I asked her how much the contract was for.

'20 million pesos [£4,000 or US$7,000]', and I asked 'How much did the cattle cost?'

'Let me see, they spent 17 or 18 million pesos on the cattle, I can't remember exactly.'

But then, since none of the other women turned up at the meeting we all went home. And after that . . . she never sent for us again or let us know about anything, nothing. We were just ignored.

So what you earn from the cattle, is it a help or haven't you made anything out of it?

No we haven't had any benefit from it so far, we don't know whether it's going to be a good thing or not.

But you have joined in looking after the cattle?

Oh yes, I've joined in the work. Quite a few of us women have put in a lot of work, we've worked on the plots, doing farmwork, clearing pastureland. Clearing, weeding, mending the fencing, but when it comes to this notion of 'co-operating' in money matters, we haven't co-operated at all, because after the extension officer's[12] visit, when they did some accounts for us and everything and there was some money left over, well we thought if we all join forces, you know, and get an agreement that this money that was left over, well it should go to pay off the debt, because our share was about 100,000 to 200,000 pesos, and we thought we'd never manage to pay it because my daughters, they're in this credit thing with the UAIM. Well, where are they going to get the money? Because, there are three of us in the UAIM.

Now that you know what kind of work the UAIM does and the work you do at home, which do you think is more worthwhile? For a woman?

Weeelll . . . I think what I do at home really, because it all goes to benefit my kids. Although the UAIM, well it does benefit all us women, it's there so that we can all help ourselves and we all ought to have . . . a good attitude to work, be willing to participate in all kinds of work.

And are you happy going to the UAIM?

Well some days I don't feel like going and other days I'm really keen to work in the UAIM, for the same reason really, because I think working like this we waste our time a bit, we get chatting and . . . well, sometimes I wonder 'Are we ever going to make any money out of the UAIM?' Perhaps we're even going to be saddled with a debt we can't pay off and we're going to create more problems for ourselves, more 'How are we going to find the money for this?' But then, well, that's how it is, sometimes the others are feeling optimistic and sometimes they're really down, because of the executive. Because the president doesn't behave properly [. . .]

And what advice would you give to other women like yourself, your age, who also have lots of children? What advice would you give them?

I'd advise them to . . . about my experiences, what I've been through. I'd advise them to do what I do. If they had . . . I mean in my opinion, other sisters[13] probably think like me and feel like me. They give advice to their daughters, where work is concerned we all ought to do the same amount, we shouldn't go around fighting with one another, we ought to get on well together, to live in peace, to be happy, or not always happy but at least peacefully! Not to be worrying about 'So and so said such and such' or 'I can't go out because she did this, or because of that' but for everyone to get on well [. . .][14]

Is there anything you would like to tell us about the life you live now?

Weeelll . . . about my life now, well I think I've told you just about everything. About my life now, let's see. Well not really, just that now that's how my life is.

Life story collected by Silvana Pacheco and Elia Pérez, transcribed by Silvana Pacheco.

NOTES

1 Elena's voice slides between herself as child and herself as mother here, explaining that mothers like to hit children.
2 El Arroyo is a hotbed of gossip; Elena's story repeatedly speaks against it.
3 A very characteristic exclamation.
4 *Curandero.*
5 To be with someone: used euphemistically for sex and implying the inevitability of sex if girl and boy are together.
6 The idea of shame is very powerful and links to 'honour and shame' in Spain (Pitt-Rivers 1971).
7 Spanish here says 'yourself', rather than 'your clothes'.
8 Fried turnovers, dumplings.
9 See note 1.
10 See note 2.
11 She tells the story of one woman who miscalled her daughter and came round drunk, with a machete, trying to break into the house.
12 *Promotora agraria*: a woman agrarian development worker.
13 Literally 'women comrades'.
14 Gossip again.

10

CLARA'S LIFE STORY

Clara was born in Cuitlahuac, in the same state as La Villa. She is 36 years old and has five brothers and two sisters.

When you were little, did you help your mother with the household chores?

I was just 6 when I started. I helped with the work all right. I used to work on a bench . . . grinding [maize], because my family was very poor, they were very poor, they used to work . . . picking coffee, that's what life was like then. Out picking coffee while we had to stay behind helping out in the kitchen, grinding [maize], making tortillas, fetching water and everything.

And then you went to school?

[. . .] So . . . I finished sixth grade, then, I went to school over there, in Otzaloton. I only did the sixth grade there. Then my teacher said to me, if I wanted, I could go on studying, but no, we couldn't do it . . . because we didn't have enough money. I said to my father, 'You know what, dad', I said to him, 'The teacher told me I can go on with my studies because I got good marks.'

'Now, hold on a minute, my dear, I can't really . . . pay for you to study. You know very well that I'm messed up, my hands are crippled and I can't earn a living. And what you have to pay for school [. . .][1]

When you were little, how did your parents treat you?

My father was so good to me, it was my mother who was nasty. She used to punish me all the time. If I was a tiny bit late, she would beat me on the spot. She was always beating me! Not my father. He was so good to me, he never beat me. It's always mothers who punish children the most, for disobeying sometimes . . . [Laughs.] That's how it is!

And what did you like best about your life as a child?

About my life as a child, well, in those days they didn't even let us play! We were, like the saying goes, 'little animals: just work and more work'. They wouldn't let us play with our friends. If we were at a loose end they would give us work to do. We hardly ever managed to get away and go out to play [. . .]

Do you remember how old you were when your periods started?

Fourteen, I was 14 years old when my periods started. I didn't know what it was, in those days no mother would tell you, 'You know what, daughter, this is going to happen to you one day, like this.'

Nothing like that, you had to find out for yourself. So suddenly it started and I didn't know why this had happened to me. I was sort of frightened, like a frightened little bird . . . Why had this happened to me? [. . .] Because even mothers and daughters didn't discuss things in those days, not even when that happened to you, the only communication was that they didn't communicate anything, anyway . . . then I began to have it month after month and I still didn't know what it was.

And what advice did your mother give you at that age?

Well, she told me to be very careful, that now I was this old I shouldn't fall in love too quickly.

Because it's true, at that age you go through a very intense phase, you fall madly in love with someone, with anyone who comes along. But she told me not to, not to accept anyone [marriage?], because I was too young, and not to have relations with anyone when I was like this because you could get pregnant. And she really must have got me worried, because I did what she said.

Did you enjoy going to dances?

Oh! I lived for it! I loved it. And it was my mum who used to take me, my mum would say to me, 'Let's go to the dance', when I was 14 she started taking me to dances. 'Come on, let's go to the dance, my dear, because one day you'll get married, and although I pray to God that you won't end up with a man who doesn't know how to take you out for a good time, at least you'll have had some good times with me'.

That's what she used to say. So she used to take me, from the minute the dance began right to the very end, it was wonderful! Yes, I remember it now. Because even though she could be nasty, my mum, she took me out to things to enjoy myself, to dances, because she wanted to look after me [laughing as she speaks]. That's how it goes [. . .]

And how did your husband treat you before you got pregnant?

Well, I mean, he did love me at the time, but then he had his own life. He was living with another woman. My life was pretty unhappy the first year we were married [. . .] Then there was the time there was a dance on and he hit me . . . He slapped me across the face, but then he was ready to carry on, he wanted to take me home. But we were at the dance and my mother said to me 'No, you're coming home with us because he's going to beat you, let's go home and settle things with him tomorrow.'

So you had gone to the dance with your mother?

Well it was like this, my husband gave me permission to go, but he was so jealous you couldn't even talk to your own family. Well, then my *compadre* [old friend]² said [to him], '*Compadre*, do I have your permission to dance with your wife, with my *comadre?*'

'Of course', he says, 'Here she is, go and dance with her.'

So he gave his permission, but I should have guessed what would happen, because he was already in a bad mood and he was drinking a lot at the dance. So then a boy comes up and says to me 'Auntie, you have to go home. My uncle says he's waiting for you.'

'But, why?'

'I don't know, but he's furious, he's waiting for you, he says he's going to beat you.'

'Well, in that case, I'm coming', I said.

So I went over to him and I wanted to say a few words to him, 'Now come on, you gave him permission . . . he asked your permission. Why are you playing these games? Because you don't mess around with a *compadre*. Besides, I wasn't even chatting to him, or flirting or anything, I was just dancing like people do, all perfectly formal', I said to him.

'Oh no you weren't, you were . . . flirting with him, you were chatting him up.'

'No I wasn't . . .'

Well, it was useless, he tried to hit me with a stick and then my mother pulled me away. I went home with her. I woke up there and I didn't want to go back to him . . . never again [. . .] So then my mother said, 'Right then, you're going to go over there, but we're going to do something. Let me have a word with him and let him know that if he intends to go on mistreating you then you'd best not go back. If he agrees, you can stay at home, in my home, after all you'd be the only woman [daughter] at home', she said. 'I accepted him as a man, I gave him a wife and I never thought that he would treat you like this.'

My brothers were really angry and they wanted to give him a good hiding, they were going to beat him up, 'Let's teach him a lesson once and for all', but then my mother said, 'No, don't touch him. Perhaps he was drunk, he had one too many. We'll let it go this time. But if it happens again, he won't get off so lightly.'

And did you want to go back to him?

No way! I wanted to stay at home! But then his mother said that I should forgive him the first time, but if it happened again, if he did it a second time, 'I'd tell you to leave myself, because, what right has he to mistreat you, if he himself gave you permission to dance, and if he's going to mistreat you, even I'm not going to put up with that.'

So then, um . . . oh yes, he apologised, yes, he was drunk when he did it. His *compadre* came by and he apologised to him, too.

'You shouldn't behave like that *compadre*, mistreating my *comadre* like you did.'

'I'm sorry if I've offended you', he said, 'I had one too many.'

But oh no, he knew what he was doing [. . .]

And when you got pregnant, were you pleased?

Oh yes! . . . I was really pleased! Really happy! Because when a child comes into the world, it's not just because we want it, but because you'd be happy with either [a boy or a girl]. If it's a boy, he's [the husband] no way of knowing. As a mother and father, you have the right to love them and give them an education and everything they need [. . .]

What was life like for you in the old village of La Villa, before it was moved by the Plan?

Well everything was fine, not . . . I didn't have anything against the place, everything was fine. Because I was happy in my work, it wasn't like here where you're shut up in your little plot of land . . . no, over there was land for rearing animals, pigs, chickens, I had plenty of animals! I didn't have to worry each year about having enough to eat, if I was hungry, I just killed an animal. But not here, you can't have them here because everything has to go to the cattle-pen. And in a cattle-pen with no pasture, what can you rear and what can you do to help? Nothing, there's nothing.

Did you like living in the old La Villa?

Yes, I liked it a lot, because it's a beautiful place! And it didn't get floods, the river never came up. It was quite high up, so you could plant a few things or whole plots behind your house.

And . . . you could grow anything there! Because that's where I used to work, behind the house we had beans and they used to give a good yield. We used to plant tomatoes and it was the same story, pumpkin . . . It all helped the man of the house because we didn't have to buy any food, we just had to reach out and pick it. It's a different story here, everything's based on money. If we have some money we can buy something and if we don't, we can't buy a thing, that's it. It's a real problem.

When you were living in the old La Villa, did you get to see the forest, all the trees and so on?

Yes, I saw it all. When we arrived, it was just the forest . . . there was no village. People were living in the reed-beds, there were no houses, there was hardly anyone there. Those were virgin forests when we got here. There was plenty to eat, all kinds of things: *tepezcuitle* [big rodents], deer, what they call armadillo – you could get all kinds of things! But as time went by, it all came to an end, because people started coming in droves and it all got used up. They were chopping it all down to make pasture. When the *ejido* came in here they destroyed all the forests to plant maize. The whole thing became one big pasture! So all the poor little animals went further and further away, they travelled a long way too, because wherever you cut down forest, everything that belongs to the forest goes away, too – the animals . . . [. . .]

How do you spend your life, to live or to survive?

Well, at first I had an electric mill, that is, I have a mill which we saved up for. That's how I managed to pay for my oldest child's education, because things weren't so hard thanks to my work with the mill. Every day I would give him his 1,000 pesos for a drink and a pasty and everyone was happy! But later on I had to give up work, because after the operation I felt awful, uh . . . there was a pain here [abdomen] and in my bones. I don't know whether it was because I went back to work so soon after giving birth to my son. I went straight back to work, cleaning maize, making *pozol* [a drink of maize and water] to sell, and so on.

Weeelll . . . I felt terrible! So then my husband said I should stop working because perhaps it was doing me harm. So I stopped working and I haven't used the mill since. But the boy who's just started in second grade, I can see that he's suffering for it, because some days I've enough to give him the money he needs and other days I haven't even enough for a drink. So I really feel bad about that, because he wants his breakfast so that he can understand what they tell him at school, because if you go without breakfast, it's much harder to take things in, and he can't pay attention to what the teacher's saying either.

So you used to grind maize, and then what would you do?

Prepare the meal! Grind the maize and then prepare the meal and in the evening . . . do the washing, the boys' clothes. That's how I used to work, I didn't have a moment's rest, from 6 in the morning until 1 in the afternoon on the mill, then I go into my kitchen to prepare lunch and as soon as everyone's finished eating, I get out my tub to do the washing and . . . that was my life. Nothing but work! Maybe that's why I got fed up, because I never got a break. It was too much work [. . .]

You said you used to sell things? What did you sell?

I used to sell pork crackling, pasties, popcorn. That was part of my work, too, to help my children, to buy them pencils and jotters, and it all came to an end. I didn't want any of it. I got fed up. Now, since then, I haven't sold anything, so there's no money for extras. If there was some maize I would go on working, but I don't have anything now, there's nothing. We've been struggling by with some cattle my husband has, but it's not much. We can barely afford to eat, but we can't afford what the school costs nowadays. We've got five of them at school and two in nursery school! If you only knew how it upsets me that I haven't enough to pay for everything, to get them the things they ask for. If I pay for food there's nothing left for the things they ask me for. It's been a hard year all right!

And what work does your husband do?

He works in the fields. He used to work in the collective [under the Plan], it got credit [without collateral] but it was nothing but trouble. Endless work! There were no Sundays any more, the weeks just ran into one another. They never got a break . . . [. . .] So he left the collective, actually they kicked him out, he didn't want to go. They kicked him out because he had a few cattle, so they said he couldn't work for credit [. . .]

And don't you help your husband in the fields now, on the plot?

Weell . . . he doesn't take me because when school's on I have to get breakfast for the children. I have the washing to do, meals to prepare.
    'Better stay here', he says, 'I'll do it' [. . .] But I do tell him, 'Just say the word, dear. I want to go and work with you, I want to help you.'

How does your husband treat you now?

Well, up to now, thanks to God, everything's fine. He's no saint . . . but he does appreciate me, despite all the years I've suffered. Because when we arrived here I had a miserable time for a while, because he likes 'the bad life'. He likes to have a good time. Once . . . I really spelled things out for him then, because there was a time when he nearly strangled me, he was drinking so much, always drinking and drinking, he really liked to drink [. . .] [He used to accuse her of having lovers.]
    'No, I'm not that kind of woman. I may be poor, my father brought me up in humble circumstances, I may be humble but I have my honour', I told him. 'Not like you. I may be poor, but you, you're throwing away your father's advice on the rubbish heap', I said to him and we went on like that. When he tried to strangle me, he got me with his machete, so then I said to him, 'Now look here, stop it. If that's how it is: it's gone far enough. You stay here with your children and I'll go without my children. I came without children and I'll go back to my home without children, because it's your duty to support your children, not for me to support them', I told him. 'Thanks to the bad life you lead, you're going to end up a bachelor again, but I'm not taking a single one of my children with me. I'm sorry, but I can't stay with you with the [bad] life you lead!' So then he begged me to forgive him, 'Don't go, I won't do it again.'

And then a group came from the church to counsel him and they told him he shouldn't live like that. 'Your family respect you, they love you, they won't see you go hungry. She tries to make a living, she goes out to work, she manages to provide you with food, and you don't ask yourself, where did she get the chicken, where did she get the food? By leading a good life, by being pleasant, she managed to get all this. Other families are happy to help her, the neighbours give her things, and when you're eating it, you never ask yourself, how did the food get on the table? What more do you want? You have an excellent wife who helps you with your work. And you're still going to carry on mistreating her?' 'No, I'm not', he said. And ever since then, thanks to God, I may have lived humbly, we've been poor, but . . . no, he's never treated me as badly as that again [. . .]³

And do you enjoy chatting with your children?

Yes, I talk a lot to my kids, but to give them advice, even though they're growing up now, I still tell them off. I tell them, 'I won't be leaving you a big inheritance and I can't give you a luxurious house, but from my experience and what little I've learned, I'm giving you what I can' [. . .]

You were telling us that you're in the UAIM working with the other women. What do you do there?

Well, it's like this . . . the way we're working there at the moment, sometimes they just send us to check the cattle when . . . it's really easy, giving them some salt, counting them. Because the animals are really wild! [. . .] [Once a month the cattle have to be properly checked over.] I let him [her husband] know. It only takes a couple of hours. I don't do it, because he's the one who handles the horse. He just goes for a little while so I don't bother going. Only when there's something easy to do, or to see how things are going, or when someone comes out from the bank, well then the women members have a meeting and you must be there. So that's when we [women] go. But anything to do with the cattle, he's always helped us, he comes to help me [. . .]

And do you like having [female] friends?

Yes, I have quite a few friends, and when people see that I'm easy to talk to, because I get on with everyone . . . In all the *ejidos*, I'm the one they talk to most, I'm the one they have a chat with [. . .]

And does your husband let you have [female] friends?

Yes. He's never unsociable with people and he lets me have a chat. Since he knows I'm the type of woman who doesn't want to have any problems with him, and that I don't go looking for trouble with anyone. Chatting to people is something . . . pleasant, when you can trust one another. Not like some people who come for a chat and start spreading idle gossip. I'm not like that. I don't like that. I like to chat to anyone who's willing to have a chat!

If a friend in the village falls ill, or one of your [female] friends, do you go to visit them?

Yes, we[4] visit them, we help them do their washing. For a while now, since we've been working together, we've been helping the old women: washing them, bathing them, looking after them, getting them something to eat. Or if we have something to spare . . . then we take it to them. And that way . . . you know what, I've got to know a lot of people, because even if your family don't like it, it's our duty to visit people. I'm Catholic[5] and I'm happy about it. I really like it, because they [the church] teach you how to behave, how to be with other people, with your [female] neighbours, how to make friends. Not with gossip but with deeds. They teach you what religion is all about [. . .]

[Clara is now studying secondary education under an adult education project.]

Which year of secondary school are you in now?

The first year of 'open' secondary school. I take classes at the school [. . .] It's not so easy to get work washing clothes. Casual work to earn a little money to help the kids. Unless you've got your certificate for your first grade from secondary school, you won't get a job. Because lots of people say, 'There's no work for poor folk. They're not capable of doing the job.'

I want to work at something else, something where I can earn a bit, even if it's not much, to get things for the children. So I don't have to keep asking my husband. So that *I* can buy things too, by myself. Through my own efforts! [. . .]

A little while ago you mentioned that you had had an operation. What was it for?

Ah, yes. It was to stop having children. Actually, I didn't want to be operated but my husband said, 'Enough is enough.' And I have to say, perhaps he was right? I mean, the situation is so tough just now. We're really struggling.[6] We can't buy enough to eat with so little money! Ten thousand pesos! Since *he* won't look for a job or anything, he just lives it up for a while whenever he sells a calf. So I suppose he did the right thing getting me operated, because now we don't have to worry about the baby being ill, really ill . . . none of that. We can relax a bit, we just have to think about the ones who are growing up now and want their meals, their money for school.

'I need a pencil. I need a jotter.'

They're always asking for something. So it's not the little one who comes crying to me now, it's the big ones. So I do feel a bit freer without the responsibility of little ones. I *did* want more children, but the way things are, it just isn't possible. There are eight of them as it is! So I started to reconsider myself. I had enough to cope with already, with what the eldest asked for: he needs new clothes, the school uniform, shoes, this and that. So what am I going to do when the next one goes to school? I said to myself. And what if I have a little one and he's hungry. Where am I going to get milk for the other one? And that's how I started to think that it was for the best that I'd had the operation [. . .]

And what advice do you give your eldest daughter?[7]

I advise my daughter not to take the wrong path in life, but to learn how to work. Learn how to work, my daughter [. . .] 'If you have your studies, if you know how to work, you'll be able to get enough to eat wherever you end up', I tell her. That's what I tell the little one.

And what advice do you give your older children?

I advise them not to be led astray. To go out with their friends but to avoid the ones who like a drink, or like to smoke marijuana, because you even get that here! There are lots of young lads, right, because there have been a lot of soldiers coming here [near the Guatemalan border]. So that stuff was coming in here from the beginning [. . .] So I tell my sons, 'I'm not having it', I tell them, 'I don't want you lot losing your way one of these days, after all the advice I've given you. Just throwing it on the rubbish heap, as if it was worthless. You must learn some discipline, you must understand what advice is for. I'm telling you now. I grew up poor, but you've never seen me wandering from pillar to post with my suitcases, lugging my things around. And why not? Because, poor as they were, my parents gave me an education and I'm making good use of it' [. . .]

Have you ever thought about what you would do if you were widowed some day?

Well, I . . . I've thought, well, what can I do? Just go on working, with whatever my husband leaves me, the little he leaves me, I'll use it to provide for my children. And I'll have to work like I was both father and mother of the household, look out for my children, make sure that I don't lose one or other of them. Boys and girls alike. That's what I've talked to them about a lot, too. That if the day comes when I can't manage any more, if I have my children who care about me and love me, well then I'll live with him [my son], if his wife lets me, and if not, well I'll stay in my own house. But I don't want them kicking me out. Throwing me away like a piece of worthless rubbish [. . .] You know, I even had a young lad coming to ask me for advice. He doesn't have a mother, just his father and stepmother. His mother was advising him to give up school to

175

work on the *parcela*, to give up his studies. She even said, 'What are you doing studying. You're just wasting your time and throwing yourself away. Soon you'll get married and that'll be the end of it.' So he comes to the house crying, because he's friends with my son, and do you know what he said to me? 'What do *you* think?, because I'm not happy about it', he says. So I asked him why, and he says, 'You're the mother of my friend Samuel, what do you think?'

'Well, what I think is that your mother's giving you bad advice. Instead of telling you, "Come on, son, you must go on with your studies. If you want to study, then do it." The land is there for those who don't want to study, but if you want to study, then carry on! Whatever the cost. Even if you go hungry sometimes, even if you go without! If you really want to go on studying, if you want to be someone in life', I told him. 'Then you have to ignore the advice she's giving you, because it won't get you anywhere.'

Well, one day he came to see me and he said, 'You were right, Clara. And you've given me something money can't buy! I've got some money now, I've got a grant, and I'm going to the *preparatoria* now [pre-university classes, like a sixth form college]' [. . .]

That's what you need in a mother. Even my own husband didn't want to go to secondary school, but then he's never listened to the advice that's given to his own children. Even I've given them advice, I've been advising them. And he's had to reconsider a few things. Like, I'm worth much more than, like I say to him, 'I may not be pretty or rich, but my experience is worth a lot. Both to the children and to you.' I'm a bit like a teacher for him.

And what advice do you, or would you, give to women of your age?

Well, I advise them not to go astray.[8] To set a good example to their children. Give them good advice, send them to school. 'If he wants to study, don't take him out of school, because it'll end in tears when you see him . . . drinking, hanging around with his friends smoking rubbish. And that's not all. The one who'll be crying is you, not him' [. . .]

You know, around here there are a lot of women who get shouted at once and they're off, 'I'm moving to the city. My husband's not going to behave like that.' They don't have any self-discipline[9], they're ignorant. They think it's easy to move to the city. Just because I've had a fight with my husband, or he threw me out, I'm not going to find work so easily if I don't have any skills. If my husband told me off for not doing my work, if he hit me, and off I go, well, working in the city is worse. It's much stricter, 'Clean the house, wash the walls, make the bed. It all has to be done a particular way and if you don't know how? *Then* how are you going to get work?' I tell them. 'After a while, the mistress of the house says to you "You don't know how to do anything properly." And then what are they going to do with you? Sack you' [. . .]

And as a woman, what do you think of the tequio[10] *in this community?*

Well, I think it's really bad, because you have to do the *ejido* work for free, without earning a penny. I mean, they may well say, 'In due course' – that's it – 'In due course, it will yield 30,000 to 40,000 pesos profit' [US$10 to $13, £6 to £10]. Well, that's in due course. What use is that to someone with a lot of kids? Maybe if they were just 1 or 2, everything would be in your favour with so few of you. But what about eight children and two adults, like us? There are ten of us and we can't make ends meet. How can we manage to go on giving free labour? [. . .] The problem is, there's no work around here. There's nothing! No way to make a living! I work hard enough, but it's all for free, *tequio*. One things for sure, *that* never runs out.

And does your husband also have to do tequio *for the* ejido?

Well, yes, even though he's not a member of the credit scheme, he still has to do his share of the *tequio* for the *ejido*. The *ejido* maize has to be fertilised, the *ejido* fertiliser has to be spread, the *ejido* weeding has to be done, the *ejido* maize has to be planted. When the harvest's ready, the *ejido* maize has to be harvested. If the *ejido* has cattle, the cattle have to be tended, the pasture has to be fenced – cut the wire, put up the posts . . . the lot.

And how do you think the problem of the tequio *could be resolved?*

Well you've got me there, I don't know what to say. Because it's all we've got around here. Sometimes, even if he wanted some work as a day labourer, there's none to be had, none. Lots of people go looking for work. They leave their work here to go and look for work somewhere else, so they can support their family for a day or two.[11] So the *tequio* business, I just don't know. No, we're just scratching our heads here because we really don't know how to solve this problem. Lots of people are thinking of getting out of the credit agreement because they don't want to do their share of *tequio* any more. Because their family says to them, 'Look, we can't go on living like this. We don't have this, we don't have that.' I need some soap. How can I buy it if they won't give credit in the shop? Even if they do, you still have to pay sometime. That's how tough the situation is, it's really, really hard! If there was some other work, something else we could do, just for a week even, then we could manage. But there isn't, there isn't [. . .]

Here in the ejido, *what do you see as the big problems affecting you as a woman?*

Problems with the housing. The problem is that there's nowhere to work. You want to make something of the yard and it's impossible. We've really had a go, with my husband lugging earth here to plant coriander, vegetable patches. It won't grow. Because sometimes the earth dries up, or it gets waterlogged, so that you can't grow anything decent [. . .]

Do you like to keep animals in the house?

Yes, I mean I had them in a pen, and I do like to keep animals.

Why do you like it?

Because it's a help. I don't have to buy them then. I mean, whenever I don't have any money I can get some with my own work, with my own efforts. I'm the one rearing them, looking after them. I can just grab something to eat. I don't need any money to go out and buy something. I just grab a chicken and kill it for my children and they've got something to eat. Or if I have some eggs I can sell them and if I haven't enough to sell, well, then I have some for breakfast. I've always liked keeping animals at home.

Doesn't your husband get angry if you sell them?

No, he doesn't interfere. I can sell a chicken, or a turkey, can't I? Because once it's sold, it's sold. He's not going to come and say, 'Hey, woman, why did you sell that?' No, no, he doesn't interfere because he knows that that's *my* job. I'm working here too, to help him, because we know that he needs to eat, too. Just as he works in his things and I don't interfere when he's sold something [. . .]

When you're with your husband, if you like making love and you enjoy it, do you feel good with him, when you're together?

Well, yes, it's something we both like. I mean, if I'm not feeling well, he doesn't . . . he doesn't try to seduce me, only . . . when I can. What he says is, when I can, fine, and when I can't, just leave it. So yes, we've both been happy with that. We do enjoy it.

He's never forced you?

Um, yes, when he's drunk. When he's drunk.

And what do you do?

Well . . . sometimes I . . . I might put up with it once, but no more. Because a woman isn't up to doing it so many times. Sometimes you feel ill and a bit worn out, so you say 'No more', you're not up to it. And what you do is you send him off to sleep and you don't move him, and you sleep on your own, and he sleeps on his own. Once he sobers up and he's back to normal, then it's a different story.[12] Then you know what you're doing. The thing is, when he's drunk, it's like he doesn't really know what we're doing . . . right, so that's why we've reached an understanding. When he's drunk, I say to him, 'Don't touch me. I'm going to sleep over here and you can sleep over there. When you've sobered up again, then . . . you can come over here.' [Laughs.]

And you don't quarrel about it?

No, no, thanks to God. Why should I say so, when we don't?

Life story collected and transcribed by Silvana Pacheco.

NOTES

1 Clara raises this point repeatedly in sections which we have edited out. A problem with shortening and editing life histories: repetition is tedious, but important.
2 Literally, *compadres* would be linked as god-parents: this is a pretence of kinship.
3 We have omitted a lengthy discussion of children's education.
4 Clara is asked personally (*tu*) but replies as 'we' (*nosotras*).
5 Catholic duty enabling her to leave the house?
6 The verb used is *sufrir*, 'to suffer'.
7 Clara was asked about her eldest daughter, but ended up talking about her 'little one'.
8 Could also mean 'let yourself go'.
9 Again, 'don't know how to suffer/endure' – *sufrir*.
10 *Tequio* is the immediate payment for collective work: in theory, there will be a share in the profits.
11 This is the only community in which our survey found more women than men, although others claim that men must go away to work a great deal.
12 She shifts from the general, sex, to the specific, drunken approaches.

11

GUADALUPE'S LIFE STORY

Guadalupe speaks Tzotzil and is from the highlands of Chiapas; she came to Tacaná with her second husband. Her family thinks she is about 44 years old. Her parents died about twelve years ago, but six of her nine brothers and sisters are still alive. Guadalupe told us her story in Spanish, her second language; we have not tried to render this as dialect.

As a child, did you help your mother with the chores?

Yes, I helped her in the kitchen, with the laundry, to [obscure] and . . . we had another occupation, picking coffee, we had to see to all that; that's my work from my childhood [. . .] I had to grind [corn], make tortillas, wash, iron . . . all that, from the age of 7 or 8, that was my work, and when I was 9 and 10, they started taking me to work in the fields, and out to work, too, that's what our life has been, miserable, bitterly hard, oh yes! . . . So much work but we didn't get anything to eat out of it, it doesn't leave you anything, you never have any fun, because the village is in the middle of the mountains, not even the doctors or the schoolteacher come there. We suffered a lot in my childhood, all my relatives suffered in their childhood, because we lived deep in the mountains . . . oh, we did.

What did you do in the fields?

We did – we weeded, we worked the *milpa*, we planted chillies and tomatoes, sugar cane, sweet potato, yuca [cassava], yams . . . all this was our work that we did, back where I come from. Coffee, we planted that, too, potatoes and all that.

Did you go to school?

Yes, I was there for one year, I had a teacher, I only went to school for one year, I left that same first year, I didn't do the second or third year, just the first year, I didn't understand a thing, but thanks to God – perhaps God laid His hand on me? I understand . . . a few things, spelling out the letters, I do understand a few things, I don't know much. But, thanks be to God, my children learned something, one of them finished sixth grade, 'though only so-so, but he does know a bit more than me, he does.

And how did your parents treat you when you were little, did they beat you or did they love you a lot?

Yes, they beat me a lot, Aiee! How my father beat me, he used to thrash me! Because we didn't go to work, because we were all girls, all girls, just two brothers, who went and did help my father with his work, because there were so many of us . . . well, the work didn't provide enough, so we had to help my parents with their work, that's all our life has been, we've had a bad life . . . a bitter life! Because after this work we got nothing, nothing to wear, nothing to eat, we had to find other [paid] work to get something to eat, even though we went to work and help, but we didn't get anything out of it, that's our life of suffering over there, but here . . . it's different, it's a bit better.

What did you like most as a child?

I liked . . . to learn, but there wasn't anyone to make it possible. I liked learning, embroidering, in my childhood I liked embroidering; I began embroidering all by myself, nobody taught me [. . .] I wanted someday to be able to be a teacher, I used to ask God to help me, oh yes, that was my thought that I carried in my childhood, I wanted to learn! But nobody helped me with this, nobody!

Do you remember how old you were when you started your periods?

My periods began when I was 13 and a half, yes, they did.

And did you know about it?

No, no I didn't know what this was, in fact I got a terrible shock! It even came with the shivers, this, this evil thing[1] came with fever and shivering, my bleeding began and for five days I had this bleeding; I was afraid to tell my mother, because my mother was ill then and she couldn't get up and I didn't want to have to explain it to her to find out what remedies I should take for this evil thing. I didn't know what was going on! There I was laid up in my bed and my mother too, and no doctor to consult [. . .]

So you didn't know about it, your mother never talked to you about it?

My mother never talked about it! She never talked to me anyway, and I never talked to her. I was really afraid . . . 'Why am I seeing all this blood?' I was afraid of my mother, I thought she would tell me off, who knows what I was imagining? [. . .]

So your parents told you how you had to behave, how you had to be?

Yes, they did, some things they did tell me: 'You lot, when you grow up, work, do your work like I'm showing you, because some day you'll get married, and some day you'll have to leave here, out of my control, because you'll have to go to another house. If someone comes to speak to me and ask for you, and I hand you over, and you go to another house, you won't be living with your mother any more, not with us' [. . .]

How old were you when you had your first boyfriend?

I was 18, yes, 18 [. . .]

So you married this man first?

Yes, I lived with him first, and we were married properly, he asked for me and we lived together for a year without any children. He got some doctors' medicine and gave me the medicine and then I had a child with him. That's the home I have far away over there: he, my son, is 25.

And then were you divorced, or how did you separate?

We separated because he brought another woman to live with us, my sister, and that's the one who has stayed with him up to now, and thanks be to God they've been all right, just as I have, so they have. Well, it was like this. We began to hear the word of God, because at the time we were living like an animal, let's compare it to an animal because we didn't know about God's ways and we did as we pleased; not what pleased God, but as we pleased. Well, then, people preaching the word of God came to the village and so we started to understand a few things and I realised that it wasn't right the way we were living and carrying on with two people, with two women. That's when we began to think that we weren't doing things properly, that one of us should separate and he should stay with the other one. And it was me who left [. . .]

But that man, he didn't want me to go, he cried a lot over me, because I was the first one, the two of us cried a lot, me and him, because he didn't want to let me go. But thanks to God it's nearly eighteen years now since we split up, thanks to God they treat me like a sister, that man and my sister, they treat me very well, they love me a lot, they treat me like their sister, like their mother. That's how we ended up, there was no other way [. . .]

When the child you had with your first husband was born, how did you feel, when he was born?

Huh! . . . I felt terrible! Thought I was never going to get better, my pains began when I was working in the *milpa*, when the maize was just green. I went to pick some early maize and I had to go back fast with my father, because we'd gone out to the *milpa*, and instead of a load of maize I came back with a load of pain.

There was a woman who my father went to talk to, to ask her to have a look at me, in another place which was just starting up, there was no midwife, nobody to take a look at me. This woman, perhaps she wouldn't be able to help, but my little one was breeched as well, we went looking for her and we ran about 5 km, over to an *ejido* where my husband was working, to look for this woman, and when she arrived, she sees me and she says 'Oh dear God, she's dying! And this poor young thing, it's her first time and her baby's breeched, how is she going to recover?' I'd been going for eight days by then, I'd put up with the pain for ages. 'You're a brave one, my girl! Eight days you've been like this!' [. . .]

How did your first husband treat you?

He treated me well and I treated him well, too, he really loved me and I loved him just as much, he never beat me, he always took me to the doctor, that's how I've got my health back a bit.

And your present husband, how does he treat you?

My husband treats me well, too, we never fight – honestly! He doesn't beat me, he never punishes me, we work together, if I need water he goes and fetches it, if he sees me lying down, if I'm ill with something and I can't get up, he does the cooking for me, my husband knows how to cook, he knows how to make soup, chicken, prepare tortillas, serve up the meal, to do everything for his children, he does, because he says they've had to struggle all their life since they were little. He had to work as a labourer and they would let them use a house for their needs: they washed their own clothes, they worked. That's how his life has been, for my husband, that's what he tells me. Yes it is [. . .]

We didn't have any land either, that's why we've wandered around, because there was no land. If there had been land in Palenque we'd never have left, but that's why we went to Paraiso, because we thought there was land there that we could work. But there isn't any, it was already occupied, all those places, they were only renting land to work, that's why he didn't last many years there, and after that he began to go away to look for work, he would leave me alone there, he began to go and work out in the cattle ranches, and that's how he heard that they were setting up an *ejido* and they were going to join up. But because they'd told us that only the people on the register had the right and anyone not on the register didn't, that's why we started to move around . . . spending one or two years in each *ejido*, until we found some land, until we got here, we came across this place and they gave us our plot of land.

Why did you come here?

We came here because . . . We were desperate for a little work! Because my husband was working, because our children were working at the time, because we didn't have enough to eat or drink! So when they arrived, they began to work, they cut down the forest and made their *milpa*, oh, yes, and the very things we need to eat began to grow, that's why we can grow things there now: tomatoes, beans, pumpkins, peas, whatever, whatever will grow. And here at last we have our *ejido*.

When you arrived here, was it early enough to see the forest?

We saw it all, we saw that it was good and here we stayed, oh yes. It was very green, there was plenty to eat, fruit from the trees and little creatures. But now, now they're cutting everything down, well, you can hardly see anything now. Because there are trees they're cutting down, they use them to make poles, for timber, for the cattle-pens and the houses, there are even some you can eat, and they're disappearing. Because when we arrived here, we came without maize, without food. Here, we've paid a high price, our lives, to set up this *ejido*, this has been our whole life. We've been as far as Zinaparo, Centenario, Escárcega, that's where our people go to look for work. We women stay here, our husbands dumped us here – if only they had built houses, but it was just a few shelters without walls, that's how we lived with our children [. . .]

And how do you get on with your husband and children?

I do get on well with him, thanks be to God. I get on nicely with my children, too, that's why they don't want to leave, because I treat them well, I talk to them when I don't understand Spanish, or certain words; they advise me and tell me the words, they know how to say more than me, they were brought up to speak Spanish, not like me, I don't speak very well and proper, my speech is very muddled, my children belong to Campeche.

Are you and your children happy here in this village?

Yes, I'm happy that they have work, but I've always exhorted my children to study. I would have liked one of my children to turn out an accountant, an engineer, or a primary schoolteacher or to have a degree[2] or something like that, but for my sake, my children don't want to go away, because sometimes I'm ill [. . .]

And now, do you still go out to the fields Doña Guadalupe?

Yes, yes I still go to the fields. I have *milpas* now, I have to prepare the soil, weed the *milpa* and everything and do the planting, that's my work, too [. . .]

And do you still speak Tzotzil?

I wish I didn't have to speak it again, I'm tired of hearing my own tongue[3] or language of one's own. My husband and I do speak it, my children never do now, they don't want to learn it, I am trying to teach them, but they don't want to hear it, they say it sounds horrible, that's what my children say; my boys say 'When so-and-so comes I wish you wouldn't speak like that, in your dialect, because I don't like it. Oh my God! It's just, I don't like you speaking in your language, lots of people make fun of you because of it, because you speak in your language.'

And yes, it's true, because . . . this place, there are very few comrades, very few of my people, people who understand you and respect you, because it's just that here they make fun of you, they say to you 'You're an Indian', and there are several lads and young boys who say that to me and I just say to them, 'I'm a little old Indian lady, that's why I speak like this. Well, I'm not Indian, we're Mexicans, we speak Spanish.'

And they say it to my face; I just forgive them because it doesn't make me angry, God made me this way, and if God hadn't given me this language I wouldn't have been able to talk, yes indeed, that's why I realise, it's not that I chose this destiny, or wanted to be like this, but because God gave me this destiny, so I tell them, 'Do you think I chose this destiny, to talk like this? No', I tell them, 'It's because it was God's will and he made me like this.' That's how I see myself, that's how it is [. . .]

Does your husband do other jobs to earn a living?

Yes, he works on the plot, he saws wood, he does lots of little jobs to support himself. He does, he clears [the forest], he saws down trees and hacks down the undergrowth with his machete. He also wants to make his own pasture and plant grass for the cattle, he wants to get hold of some cows to fatten them up, so do his sons, and he has the same idea as them. My husband also takes in sewing to support himself, we get something to eat out of that when the pennies are running out, he sews for other people. My sons are beginning to sew, too. This sewing thing, we thought about it too, getting the other women in the community to join in and learn, so that one day they'll come out knowing how to sew [with a machine]. The rest of us *compañeras* are thinking about it, what's to stop women copying these groups and setting up their own organisation? Like we did in the meeting of all the women, these machines will be for everyone [. . .]

What did you feel when you saw that you didn't have your periods any longer?

I'm not sure, really, the last one was pretty heavy, but since there's a clinic here now, I went to the clinic with my husband and he got me some medicine and I took it, just some little pills and that controlled it . . . I think I don't ever want to see it back again, I don't want it any more because it's pretty unpleasant.

Do you have many friends here in the village?

Yes, yes I do have friends, I really enjoy myself now. I enjoy having a good chat, singing, and sometimes I like to sing in church, I go to church, that's how I enjoy myself, because there's nowhere to enjoy yourself around here.

Which religion are you?

I'm Catholic, that's right, and I chat to the other women there, we chat inside or outside the church.

When you go to fetch water do you chat to the other women?

Yes, yes, we have a chat . . . we tell one another stories, how we're suffering, how we're dealing with our problems . . . all that kind of thing, trials and tribulations, how we get ill fetching the water, that's what they all talk about, too, the same as me. Them too, that they want to go somewhere else because they suffer a lot here because of the water, because it's a long way away. This one wants to separate from her husband because she's got problems and nobody helps. I've talked about all this with them, they all want to separate because their husbands don't help them . . . [Laughs.] [. . .] Well, I say 'Why do you want to separate? There's no point separating now that you've got children, because it's not the children's fault, but if a mother thinks she's unhappy and separates from her husband, it's the children who are going to suffer, so just as you love them from the moment they're born, you have to love them to the end of your own life', that's what I've told them [. . .]

Have you taken part in any mixed groups, of men and women?

We've set up a group of men and women, it's part of the doctrine of awareness, we've set it up here, we have a brigade, because the brigade committee wants us especially to look at, 'What are your problems? About your life, how we spend our lives' [. . .]

In this religion, do they explain how the rich treat you and how the poor are?

Yes, they tell us that, too, they tell us that the rich have and the poor have not, we go to buy something from a rich person and they're not even capable of giving you a fair price for the things we buy, whether or not you have money, they push up the prices of things by so much and you just have to pay the prices they ask. That's right, you just have to pay, because a rich person, well . . . the more money they have the more they want, and we, although we're poor, we need to dress up and feel good and look our best, too . . . Even if clothes and cloth cost a fortune, out of necessity you have to buy it, and they, the rich, aren't capable of lowering the price for you, that's how it is . . . The rich treat us badly, poor people help one another out, what another poor man has he'll even give it to you at a lower price, we see it happening here [. . .]

And now how do you see life here in the village, has it changed between your arrival and now?

Yes, yes it has changed a bit, it has changed, because it was harder before, but now, thanks be to God, because we have a road now, we can get out more easily with our kids when they fall ill, because when we first arrived there was no road, and we didn't know our way around, several little ones from here died, from the community, because there was still a lot of forest [. . .] But it seems that, thanks to God, things have changed quite a bit, because there are quite a few houses in the village now, it's changing a bit now, there's a clinic, a CONASUPO[4] where you can shop, it's changed a little, there are three days of schooling now, because the teachers come here now, only they don't do their job very well. Maybe they, sometimes I think to myself that on the one hand, they've got good reasons for not coming here, because we're short of food here, there isn't enough, and no one gets any for them, and they can't put up with hunger the way we can. Maybe that's why the teachers don't come, because sometimes we don't even have soup, or tomatoes, to eat [. . .]

Our life has also changed a bit because they've given out the land and everyone has 20 ha., every *ejidatario*. But there may be plenty of land but it hasn't been shared out fairly, I don't understand why, I don't see the benefit of it. Because the top man himself, the elected leader, called a meeting. They went to get the man from the SRA[5] to come and certify it: they had already given out the land out past Cardenas, they sent an engineer [surveyor] to mark out the boundaries and it was going to take part of Tacaná, they didn't think of that. I'm [just] a woman, my husband has suffered, we carry these things in our very bones, we've fought for this *ejido*. For whom? Not for my own husband, not for any individual, this struggle, it's for everyone here! That's what I think, that it's for all the children [. . .]

187

You fought for the land. What did you do to get it?

That's right. They went to Mexico City, walking all the way, my husband went and some other Cardenistas.[6] My son went, my husband and . . . *Hijole!* Too many to count! There were several men who walked all the way there [2,000 km], my brother and other families with their children, and there were even some old people, they went hungry to earn those bits of land [. . .]

How long ago did you begin working as a midwife?

From the age of 18. I was still single then, I hadn't got married yet; I started because there was a woman who fell ill because she was going to have her baby. She was just about to give birth, and they asked me to look after her. She says to me 'How can you help me? I can feel a pain, I want you to look after me, you must know where there's a midwife or a birth attendant here somewhere. I feel really ill.'

'We don't know anyone, there's no midwife around here, I suppose it's just not possible.'

'I just need someone to hold my hand and help me because the pain's killing me', the woman says; so I helped her, I started to massage her belly and that's how I started to bring her on, who knows where I got the idea? Perhaps God inspired me, because He loves us all, his sons, his daughters. I don't know how it happened, because I didn't know how to do it, nobody showed me, I'd never seen a birth, I didn't know what one was like, because my mother and my other relatives and the people in the village never let us go in where there were sick women giving birth, we didn't get to see when they were going to get better; so I didn't know how to do the job, I learned all by myself.[7]

And do you like working as a midwife, helping women?

Yes, I do like helping them because it's always tricky, but actually, it has always gone well for me, I've handled some difficult births, because little ones will come feet first, sideways [breech delivery], bottom first, they've stopped breathing, they're mottled, feet first, hands first, and I've healed them and helped them really well, I've fought for them, but when I really can't do anything for them because they can't come out, then I send them to the doctor, to the hospital instead, because I don't have the resources, I have to say, to do a really good job, I don't have any instruments or anything, here I am working just as I am . . . with my bare hands and easy as you go! A job which looks easy, but it's anything but! [. . .]

Do you charge the women for your work as a midwife?

No I don't charge them, I only accept whatever they can and want to give me; like I said, they're short of money too [. . .] but sometimes I get as much as

10,000 or 20,000 pesos . . . others give me 5,000 pesos [£1 or US$1.60] or whatever they can and that's . . . that's all.

Do a lot of women come to ask for your help?

Yes, when there are just a few of them, sometimes five, two, six or thereabouts, I do the work here, I have worked in other villages but my husband doesn't want me going out any more, he's very strict, he won't let me go out any more, only here in my village he lets me go out to work.

Is your husband a jealous man?

No, thank God, he's not jealous, he knows how I look after myself, how I behave, how I treat the men, how I treat the women; you ought to be responsible for yourself, try to respect others, men and women. That's how you make friends, that way nobody gets shown disrepect [. . .]

Doña Guadalupe, now that you are a mature woman, you've had your children and you're watching them grow up, what advice do you give your daughters?

Well, I advise them to prepare themselves, to get working, to think of something they can study for, whether they want to be a schoolteacher or they don't want to be a schoolteacher, perhaps a doctor or work in midwifery like me, because the day will come when they'll need it, but they have to prepare themselves in plenty of time or it might be too late. If something happens and they're left with nothing to live on, it'll be miserable for them, that's what I've told my sons, too. I tell them [her daughters] they need to know how to work really well, how to cook, to do the household chores, to grind [corn], make tortillas, sweep and sew, too, so that the day they get married, when they go to live in another house with their husbands, what they did in their house they can do there. If they end up with a man who doesn't know how to work, then 'You can look for something to support yourselves', I've told my daughters, 'I want you to turn out well, not to be wandering from pillar to post, to behave well, for husband and wife alike to live in peace, happy, without anyone going to the bad, neither him nor her', that's what I always advise the girls.

And what advice do you give your sons?

I give them the same advice, not to mistreat their wives, not to hurt them, to love them a lot, for them [the wives] to love them back, that just as they love their mother, that's how they should love their wives. That's right, that's what I've told my sons; I don't want them to pick fights with their wives, I want them to live well, to live in peace, too.

'God gives us our heads to think, so you should use it, if you can't see

189

anywhere to go out and enjoy yourselves, go to church and pray with your wives so that things go well for you; you must listen to the word of God, so that you don't go around getting into fights, like cats and dogs.'

Those animals do it because they don't know any better, but we are people and we do know better, and we have to take the path of good, as God wills [. . .] That's what I tell them.

And now, as a mature woman, how does life feel, how do you feel?

I feel life as a burden. I'm going a bit downhill, I don't feel like a young woman any more, I don't have so much strength now; it's because both my legs are bad now, like they're giving way, they're very shaky, just because I had to walk a lot one day that I went to see the doctor. I walked 30 km from here to Centenario on the way there and another 30 on the way back, so all in all that was 60 km there and back, on foot to Centenario, walking around in the middle of the night. We left here about one in the morning; we left Centenario about three in the morning to get back, we got back here about eleven or noon, so I was going pretty fast, that's why I don't feel well now. For two days now I haven't been to fetch water, yesterday and today, I can hardly walk, I feel pretty bad like this, the doctor says I just need some vitamins and some rest [. . .] Even when a birth is going well, I always have to be there, even in the middle of the night or whatever, off I go, sometimes I have to go without shoes [sandals[8]] or a wrap and that doesn't do me any good. It's like I was telling you . . . what with having to count the pennies, sometimes there isn't even enough for a pair of shoes [sandals], not even a pair of shoes! That's how it is, thank God that my husband went to Mexico City last year and he brought me two coats . . . that's right, someone gave them to him, so now I have my two coats to wear when I go out.

Have you thought what you would do about bringing up your children if you were widowed, how would you keep on living?

No way, I couldn't do it . . . I don't want to be a widow! [Laughs.] Ha! Ha! Ha! What would I do? If God takes my husband and leaves me, I won't be long in dying myself, because we're so used to being together. Listen, even when I'm just out somewhere, he gets sad! [. . .] It's different for us now that the boys don't live here anymore. Because we're always worrying about them, 'What if something happens to them?' I hope nothing happens to them on the way [. . .] because they mugged my son! When he was on his way to see me, my son came, the one who was in San Ignacio, and they beat him up on the way! They robbed him and they took everything he had! They took his suitcase with all his clothes, his money he was carrying, 100,000 pesos, it was all he had. They took his towel, his shirt, his trousers . . . everything. That's why they all worry about me and don't like me going out, in case I get mugged on the way [. . .]

190

When you are with your husband and you make love, do you enjoy it?

Yes.

Has he ever forced you?

Yes, because he wants more children and, well . . . that's it, he just pesters me, what can I do? I mean, sometimes I don't want to, because I'm a woman who really doesn't like it that much, and when I don't like him pestering me, then sometimes he gets angry, now and again. But not always, just once or twice he's been really angry, yes, that's right, at first he used to force me a lot, being a lot younger than me, he wanted it more than I did and he used to force me, and then we used to fight, because I didn't want to and he did. I explained to him that I don't like doing it much, just once or twice, now and again, that I would be happy like that, but that I didn't like being forced into it. That's what I told him, yes, but he was happier when I explained. During my illness, now [menopause], he never forced me. He doesn't go around pestering other women either. I've never had that problem, him going with other women, that's why I love him so much, and he loves me too. Thank the Lord, my husband has never been a womaniser or a villain, he's never gone off with another woman, he never thinks about marrying someone else, because we've had the religious and the civil ceremonies, we're married under both laws. He'd never dream of hurting a soul [. . .]

Now that you've stopped menstruating, what does your husband say, because he wanted more children, didn't he?

Yes, I don't think we're going to have any more, and it can't be helped, if it's just not possible, then what can we do about it?

'Perhaps the time has come when we're not going to have any more children', that's what he says to me. 'I had hoped to see a few more youngsters, but since you've done God's duty, dear Lord, well then, it can't be helped, that's the end of it'.

While I was bringing up the last child, I waited three years before trying to get pregnant again, and he was getting very upset that I wasn't getting pregnant, because he wanted more children. Now he doesn't talk about it any more.

Before you stopped menstruating, did you take the pill to stop yourself getting pregnant?

Yes, I took the pill, that's why my children are so far apart, all of them [. . .] My womb collapsed [prolapse], I'm still almost as bad now. I almost don't want to get it fixed, because with a collapsed womb I can't lift heavy things, I can't do the really hard work [. . .]

You came here with your husband. What advice would you give to other women who are setting up a village somewhere else?

I tell them – because sometimes I run into women who are beginning to set up their village – I advise them that they should fight, too, that they should work hard, organise themselves well, that the men should set up a group to work and co-ordinate things . . . That they should go to the authorities and talk to them and find out whether or not the lands are occupied. That's what I've told them, so that people don't go into it for nothing [. . .]

As a woman who has lived here since the village was first being set up, who has spent a long time here, what changes would you like to see in your village?

I'd like to change . . . for us to have things we need, electricity, mills to grind the maize and a shop with machines to make tortillas. I'd like us to have a market for selling things, I'd like us to have drinking-water for the whole community and some help [credit], because if we help someone who wants to be governor or leader of the municipality, then they ought to help us, that's a real problem. We want them to help us just as our group and our people help them, that's how we think it should be. We want to see this village properly set up, get a park next to the school so that our kids grow up more healthy, that would be a big change, there will be somewhere for young people to enjoy themselves as well as the children . . . for young people to go out and have a good time, because at the moment there's nowhere. Yes, that's what we want, a few changes to change our lives for the better.

What changes would you like to see here in the village for the sake of your children?

I'd like there to be a secondary school with proper teachers, so that my children can learn and get knowledge to that level. That way, when they knew all that and had an education, the whole village would benefit. They could benefit the place. Because once our children can read and write properly, they can get a government job, right here in their own village, they can work here for everyone's benefit. That's what I would like, that's what I have in mind. We want a school, we want them to try to help us [. . .]

What about the women here in the community, do you think they could get together to ask for these things?

I think so. If they want to [. . .] I already spoke to two of the women yesterday. They asked me why I thought you [researchers] wanted to hold this meeting. Because they don't know you, so they don't have any idea why. Do you know why I had some idea? Because when we have our Squad of Jesus meetings, we talk about material things as well as spiritual things. People talk openly. They talk about God, about His word and His good advice. So I started talking to these two women yesterday, because they asked me why your group was giving them photos, why would you be giving out all this stuff? Why have a meeting of all the women, they'd never seen anything like that before. So I advised them, I told them that you were giving us these things because you wanted to treat us like friends, like the sisters that we all are, and that's why these five sisters were here, 'You have to understand, they're not from somewhere near here, they've come from another country. Other women come who live in Tapachula, or Michoacán. We've never visited these places', that's what I tell them. 'We don't know these places, because we never offer our help. We're incapable of going to work in another country, even our children are. And why? Because there's nobody to direct us or organise us properly. That's why we're stuck here.'

That's what I told them . . . I'm no great thinker, but I can tell you this. That's what they're like. I never attended any meetings at school and I've never listened to the teachers arguing, but this is what I think. And that's what I told these women when they asked me 'Why on earth do they want this meeting?'

Life story collected and transcribed by Silvana Pacheco.

NOTES

1 Literally, 'my evil/bad thing/misfortune'.
2 Compare Chapter Nine, p. 160.
3 Literally, 'dialect': indigenous languages in Mexico are not dignified by the name of language but called dialects.
4 Low price government shop.
5 The SARH, the Ministry of Agriculture.
6 Supporters of the Democratic Revolutionary Party call themselves Cardenistas after their leader, Cuauhtémoc Cardenas.
7 Guadalupe always refers to women giving birth as 'sick' or 'ill'.
8 Sandals are worn by the very poor, particularly indigenous people.

BIBLIOGRAPHY

Alsop, R. (1993) 'Whose interests? Problems in planning for women's practical needs', *World Development* 21(3): 367–77.

Alvarez-Buylla Roces, M.E., Lazos Chavero, E. and Garcia-Barrios, J.R. (1989) 'Home-gardens of a humid tropical region in Southeast Mexico: An example of an agro-forestry cropping system in a recently established community', *Agroforestry Systems* 8: 133–56.

Apthorpe, R. (ed.) (1968) *Land Settlement and Rural Development in Eastern Africa*, Kampala: Transition Books.

Arizpe, L. (1986) 'Las mujeres campesinas y la crisis agraria en América Latina', *Nueva Antropología* 8 (30): 37–66.

Arizpe, L. and Botey, C. (1987) 'Mexican agricultural development policy and its impact on rural women', in C.D. Deere and M. Leon (eds) *Rural Women and State Policy: Feminist Perspectives on Latin American Agricultural Development*, Boulder, Colo.: Westview, pp. 67–83.

Arizpe, L., Botey, C., Salinas, F. and Velásquez, M. (1989) 'Efectos de la crisis económica 1980–1985 sobre las condiciones de vida de las mujeres campesinas en México', in UNICEF *El Ajuste Invisible: Los Efectos de la Crisis Económica en las Mujeres Pobres*, Bogotá: UNICEF, p. 247.

Arizpe, L., Botey, C., Paz, F. and Velásquez, M. (1993) *Cultura y Cambio Global: Percepciones Sociales sobre la Desforestación en la Selva Lacandona*, Mexico City: Miguel Angel Porrúa.

Armitage, S. (1987) 'Through women's eyes: A new view of the West', in S. Armitage and E. Jameson (eds) *The Women's West*, Norman, Okla. and London: University of Oklahoma Press, pp. 9–18.

Armitage, S. and Jameson, E. (eds) (1987) *The Women's West*, Norman, Okla. and London: University of Oklahoma Press.

Armstrong, A. (1987) 'Planned refugee settlement schemes: The case of the Mishamo Rural Settlement, Western Tanzania', *Land Reform, Land Settlement and Cooperatives* 1–2: 30–52.

Arndt, H.W. (1988) 'Transmigration in Indonesia', in A.S. Oberai (ed.) *Land Settlement Policies and Population Redistribution in Developing Countries*, New York: Praeger.

Bahrin, T.S., Thong, L.B. and Dorall, R.F. (1988) 'The Jengka Triangle: A report on research in progress', in W. Manshard and W.B. Morgan (eds), *Agricultural Expansion and Pioneer Settlement in the Humid Tropics*, Tokyo: United Nations University, pp. 106–16.

Bain, J. (1992)'Women and the environment in rural Mexico', unpublished M.A. thesis, University of Durham.

—— (1993) 'Mexican rural women's knowledge of the environment', *Mexican Studies/Estudious Mexicanos* 9 (2): 259–74.

Basnett, S. and Lefevere, A. (eds) (1990) *Translation, History and Culture*, London: Pinter.

Basnett-McGuire, S. (1991) *Translation Studies*, London: Routledge.

194

Basso, E. (ed.)(1990) *Native Latin American Cultures Through Their Discourse*, Bloomington, Ind.: Indiana University Press.

Bebbington, A. (eds) (1993) *Non-governmental Organizations and the State in Latin America: Rethinking Roles in Sustainable Agricultural Development*, London: Routledge.

Beenstock, M. (1980) *Health, Migration and Development*, London: Gower.

Belshaw, D.G.R. (1984) 'Planning and agrarian change in East Africa: Appropriate and inappropriate models for land settlement schemes', in T. Bayliss-Smith and S. Wanmali (eds) *Understanding Green Revolutions*, Cambridge: Cambridge University Press, pp. 270–9.

Benería, L. and Roldán, M. (1987) *The Crossroads of Class and Gender*, Chicago, Ill. and London: University of Chicago Press.

Bergquist, C., Peñaranda, R. and Sánchez, G. (eds) (1990) *Violence in Colombia: The Contemporary Crisis in Historical Perspective*, Wilmington, Del.: Scholarly Resources Inc.

Boserup, E. (1970) 'Present and potential food production in developing countries', in W. Zelinsky, L.A. Kosinski and R.M. Prothero (eds) *Geography and a Crowding World*, New York: Oxford University Press, pp. 100–13.

Boulding, E. (1983) 'Measuring women's poverty in developing countries', in M. Buviníc, M.A. Lycette and W.P. McGreevey (eds) *Women and Poverty in the Third World*, Baltimore, Md: Johns Hopkins University Press, pp. 286–300.

Bowman, I. (1931) *The Pioneer Fringe*, New York: American Geographical Society Special Publication 13.

Breines, W. and Gordon, L. (1983) 'The new scholarship on family violence', *Signs* 8 (3): 490-531.

Brown, E. (1991) 'Tribal peoples and land settlement: The effects of Philippine capitalist development on the Palawan', unpublished Ph.D. thesis, State University of New York at Binghampton.

Brunt D. (1992) *Mastering the Struggle: Gender, Actors and Agrarian Change in a Mexican Ejido*, Amsterdam: CEDLA Latin American Studies no. 64.

Bunyard, P. (1990) *The Colombian Amazon: An Update on Policies for the Protection of its Indigenous Peoples and their Environment*, Bodmin: The Ecological Press.

Burgos-Debray, E. (1984) *I, Rigoberta Menchu*, London: Verso.

Butler, J.R. (1985) 'Land, gold and farmers: Agricultural colonization and frontier expansion in the Brazilian Amazon', unpublished Ph.D. thesis, University of Florida.

Calvo, A., Garza, A.M., Paz, F. and Ruíz, J.M. (1992) *Sk'op Antzetik*, San Cristóbal de las Casas: CEI–UNACH.

Cameron, D. (1990) *The Feminist Critique of Language*, London: Routledge.

Campbell, J.C. (1991) 'Wife-battering: Cultural contexts versus Western social sciences', in D.A. Counts and J.K. Brown, *Sanctions and Sanctuary: Cross-Cultural Perspectives on the Beating of Wives*, Boulder, Colo.: Westview, pp. 229–49.

Cernea, M.M. (1991a) 'Involuntary resettlement: Social research, policy and planning', in M.M. Cernea (ed.) *Putting People First: Sociological Variables in Rural Development*, 2nd edn, New York and Oxford: Oxford University Press for the World Bank, pp. 188–218.

—— (ed.) (1991b) *Putting People First: Sociological Variables in Rural Development*, 2nd edn, New York and Oxford: Oxford University Press for the World Bank.

—— (1993a) 'Anthropological and sociological research for policy development on population resettlement', in M.M. Cernea and S.E. Guggenheim (eds) *Anthropological Approaches to Resettlement: Policy, Practice and Theory*, Boulder, Colo.: Westview, pp. 13–38.

—— (1993b) 'Disaster related refugee flows and development-caused population displacement', in M.M. Cernea and S.E. Guggenheim (eds) *Anthropological Approaches to Resettlement: Policy, Practice and Theory*, Boulder, Colo.: Westview, pp. 375–402.

Cernea, M.M. and Guggenheim, S.E. (eds) (1993) *Anthropological Approaches to Resettlement: Policy, Practice and Theory*, Boulder, Colo.: Westview.

Chambers, R. (1969) *Settlement Schemes in Tropical Africa*, London: Routledge.

—— (1983) *Rural Development: Putting the Last First*, London: Longman.

—— (1992) 'Rural appraisal: Rapid, relaxed and participatory', Brighton: *Institute of Development Studies, Discussion Paper* 311.

Chambers, R. and Moris, J. (1973) *Mwea: An Irrigated Rice Settlement in Kenya*, Munich: Weltforum Verlag.

Chambers, R., Pacey, A. and Thrupp, L. (1989) *Farmer First: Farmer Innovation and Agricultural Research*, London: Longman.

Chant, S. (1984) 'Las Olvidadas: A study of women, housing and family structure in Queretaro, Mexico', unpublished Ph.D. thesis, University of London.

Chole, E. and Mulat, T. (1988) 'Land settlement in Ethiopia', in A.S. Oberai (ed.) *Land Settlement Policies and Population Redistribution in Developing Countries*, New York: Praeger, pp. 168–201.

Clifford, J. (1986) 'Introduction: Partial truths', in J. Clifford and G. Marcus (eds) *Writing Culture: The Poetics and Politics of Ethnography*, Berkeley, Calif.: University of California Press.

Coberly, R. (1980) 'Maternal and marital dyads in a Mexican town', *Ethnology* 19(4): 447–57.

Colombia (1986) *Colombia: XV Censo Nacional de Población y IV de Vivienda*, Bogotá: Departmento Administrativo Nacional de Estadistica.

Colson, E. (1960) *The Social Organisation of the Gwembe Tenga*, vol. 1: *The Human Consequences of Resettlement*, Manchester: University of Manchester Press.

Cotterill, P. (1992) 'Interviewing women: Issues of friendship, vulnerability and power', *Women's Studies International Forum* 15 (5–6): 593–606.

Counts, D.A. and Brown, J.K. (eds) (1991) *Sanctions and Sanctuary: Cross-Cultural Perspectives on the Beating of Wives*, Boulder, Colo.: Westview.

Court, G. (1986) 'The socioeconomic context of prostitution in contemporary Latin America', unpublished M.A. thesis, University of California, Los Angeles.

Cubitt, T. (1988) *Latin American Society*, New York: Longman.

Currier, R. (1966) 'The hot cold syndrome and symbolic balance in Mexican and Spanish American folk medicine', *Ethnology* 5: 251–63.

Davies, C.B. (1992) 'Collaboration and the ordering imperative in life story production', in J. Watson and S. Smith (eds) *De/Colonizing the Subject*, Minneapolis, Minn.: University of Minnesota Press, pp. 3–19.

Deere, C.D. (1987) 'The Latin American agrarian reform experience', in C.D. Deere and M. León, *Rural Women and State Policy: Feminist Perspectives on Latin American Agricultural Development*, Boulder, Colo.: Westview.

Deere, C.D. and León de Leal, M. (1982) *Women in Andean Agriculture*, Geneva: ILO.

Díaz Guerrero, R. (1974) 'La mujer y las premisas histórico-socioculturales de la familia mexicana', *Revista Latina de Psicología* 6: 7–16

Dorst, J. (1987) 'Rereading *Mules and Men*: Towards the death of the ethnographer', *Cultural Anthropology* 2 (3): 305–18.

Douglas, C.B. (1984) 'Toro muerto, vaca es: An interpretation of the Spanish American bullfight', *American Ethnologist* 11 (2): 242–58.

Dozier, C.L. (1969) *Land Development and Colonization in Latin America: Case Studies of Peru, Bolivia and Mexico*, New York: Praeger.

Ewell, P.T. and Poleman, T.T. (1980) *Uxpanapa: Agricultural Development in the Mexican Tropics*, New York: Praeger.

Fairbanks, C.L. (1983) 'Garmented with space: American and Canadian prairie women's fiction', unpublished Ph.D. thesis, University of Minnesota.

Fals Borda, O. (1986) *Historia Doble de la Costa*, vol. 4: *Retorno a la Tierra*, Bogotá: Carlos Valencia Editores.

Farmer, B. (1957) *Pioneer Peasant Colonization in Ceylon*, London: Oxford University Press.

Feder, E. (1982) 'Lean cows, fat ranchers', mimeo.

Fernandez Serra, M.T. (1993) 'Resettlement planning in the Brazilian Power Sector: Recent changes in approach', in M.M. Cernea and S.E. Guggenheim (eds) *Anthropological Approaches to Resettlement: Policy, Practice and Theory*, Boulder, Colo.: Westview, pp. 63–86.

Foord, J. and Gregson, N. (1986) 'Patriarchy: Towards a reconceptualisation', *Antipode* 8 (3).

Frick, M.J.B. (1982) 'Women writers along the rivers: The roles and images of women in Northwestern Missouri and Northeastern Kansas as evidenced by their writings', unpublished Ph.D. thesis, University of Missouri.

Fromm, E. and Maccoby, M. (1970) *Social Character in a Mexican Village: A Sociopsychoanalytic Study*, Englewood Cliffs, NJ: Prentice Hall.

Gluck, S.B. (1991) 'Advocacy oral history: Palestinian women in resistance', in S.B. Gluck and D. Patai (eds) *Women's Words: The Feminist Practice of Oral History*, London: Routledge, pp. 205–20.

Goetz, A.-M. (1991) 'Feminism and the claim to know: Contradictions in feminist approaches to development', in R. Grant and K. Newland (eds) *Gender and International Relations*, Milton Keynes: Open University Press, pp. 133–57.

Gomez Pompa, A. (1990) 'El problema de la deforestación en el trópico húmedo mexicano', in E. Leff (ed.) *Medio Ambiente y Desarrollo en México*, Mexico City: UNAM.

Gonzalez de la Rocha, M. (1986) *Los Recursos de la Pobreza: Familias de Bajos Ingresos de Guadalajara*, Guadalajara, Mexico: El Colegio de Jalisco.

González Montez, S. and Iracheta Cenegorta, P. (1967) 'La violencia en la vida de las mujeres campesinas: El Distrito de Tenango, 1880–1910', in C. Ramos (ed) *Presencia y Transparencia: La Mujer en la Historia de México*, Mexico City: El Colegio de México, pp. 111–40.

González Pacheco, C. (1983) *Capital Extranjero en la Selva de Chiapas*, Mexico City: UNAM.

Gosling, L.A.P. and Abdullah, H. (1979) 'Rural population redistribution', in L.A.P. Gosling and L.Y.C. Lim (eds) *Population Redistribution: Patterns, Policies and Prospects*, New York: United Nations Fund for Population Activities.

Greenberg, J.B. (1989) *Blood Ties: Life and Violence in Rural Mexico*, Tucson, Ariz.: University of Arizona Press.

Guggenheim, S.E. (1993) 'Peasants, planners and participation: Resettlement in Mexico', in M.M. Cernea and S.E. Guggenheim (eds) *Anthropological Approaches to Resettlement: Policy, Practice and Theory*, Boulder, Colo.: Westview, pp. 201–28.

Guinness, P. (ed.) (1977) *Transmigrants in South Kalimantan and South Sulawesi*, Yogyakarta, Indonesia: Population Institute, Report 15.

Hahn, N.D. (1982) 'Women's access to land', *Land Reform, Land Settlement and Cooperatives* 1–2: 1–11.

Hamilton, S. (1986) 'An unsettling experience: Women's migration to the San Julian Colonization Project', Institute for Development Anthropology, Inc., Working Paper 26.

Hanger, J. and Moris, J. (1973) 'Women and the household economy', in R. Chambers and J. Moris (eds) *Mwea: An Irrigated Rice Settlement in Kenya*, Munich: Weltforum Verlag, pp. 209–44.

Haraway, D.J. (1988) 'Situated knowledges: The science question in feminism as a site of discourse on the privilege of partial perspective', *Feminist Studies* 14: 575–99.

—— (1991) *Simians, Cyborgs and Women: The Reinvention of Nature*, London: Free Association Books.

Harding, S. (1991) *Whose Science: Whose Knowledge? Thinking from Women's Lives*, New York: Cornell University Press.

Harris, K. (1983) 'Women and families on Northeastern Colorado homesteads', unpublished Ph.D. thesis, University of Colorado at Boulder.

——— (1987) 'Homesteading in Northeastern Colorado, 1873–1920: Sex roles and women's experience', in S. Armitage and E. Jameson (eds) *The Women's West*, Norman, Okla. and London: University of Oklahoma Press, pp. 165–78.

Hecht, S.B. (1985) 'The Latin American livestock sector and its potential impacts on women', in J. Monson and M. Kalb (eds) *Women as Food Producers in Developing Countries*, Los Angeles, Calif.: University of California Press.

Hecht, S.B. and Cockburn, A. (1990) *The Fate of the Forest*, London: Penguin.

Higgs, J. (1978) 'Land settlement in Africa and the Near East: Some recent experience', *Land Reform, Land Settlement and Cooperatives* 2: 1–24.

Hirschman, A.O. (1963) *Journeys Towards Progress*, New York: 20th Century Fund.

Hoover, H. (1930) Radio address, 22 April, cited in I. Bowman (1931) *The Pioneer Fringe*, New York: American Geographical Society Special Publication 13.

Hulme, D. (1987) 'State-sponsored land settlement policies: Theory and practice', *Development and Change* 18: 418–36.

——— (1988) 'Land settlement schemes and rural development: A review article', *Sociologia Ruralis* 28 (1): 42–61.

Jackson, C. (1985) *The Kano River Irrigation Project*, West Hartford, Conn.: Kumarian Press.

Jacobs, S. (1989) 'Gender relations and land resettlement in Zimbabwe', unpublished D.Phil. dissertation, University of Sussex.

James, W.E. (1979) 'An economic analysis of public land settlement alternatives in the Philippines', unpublished Ph.D. thesis, University of Hawaii.

——— (1983) 'Settler selection and land settlement alternatives: New evidence for the Philippines', *Economic Development and Cultural Change* 31 (3): 571–86.

Jameson, E. (1982) 'May and me: Relationships with informants and the community', in E. Jameson, (ed.) *Insider/Outsider Relationships with Informants*, SIROW Working Paper 13, Tuscon, Ariz. University of Arizona.

——— (1987) 'Women as workers, women as civilisers: True womanhood in the American West', in S. Armitage and E. Jameson (eds) *The Women's West*, Norman, Okla. and London: University of Oklahoma Press, pp. 145–64.

Jeffery, J.R. (1979) *Frontier Women in the TransMississippi West 1840–1880*, New York: Hill & Wong.

Jensen, J.M. and Miller, D.A. (1980) 'The gentle tamers: New approaches to the history of women in the American West', *Pacific Historical Review* 49: 173–213.

Joly, L.G. (1989) 'The conversion of rain forests to pastures in Panama', in D.A. Schumann and W.L. Partridge (eds) *Human Ecology of Tropical Land Settlement in Latin America*, Boulder, Colo.: Westview Press, pp. 86–130.

Jones, J.J. (1990) *Colonization and Environment: Land Settlement Projects in Central America*, Tokyo: United Nations University Press.

Katz, C. (1992) 'All the world is staged: Intellectuals and the projects of ethnography', *Environment and Planning D: Society and Space* 10: 495–510.

Katzman, M.T. (1978) 'Colonization as an approach to rural development: Northern Paraná, Brazil', *Economic Development and Cultural Change* 26: 709–25.

Kearney, R.N. and Miller, B.D. (1983) 'Sex-differential patterns of internal migration in Sri Lanka', *Peasant Studies* 10 (4): 223–50.

Kedar, L. (ed.)(1987) *Power Through Discourse*, New Jersey: Ablex.

Kerns, V. (1991) 'Preventing violence against women: A Central American case', in D.A. Counts and J.K. Brown (eds) *Sanctions and Sanctuary: Cross-cultural Perspectives on the Beating of Wives*, Boulder, Colo.: Westview, pp. 125–38.

Kohl, S.B. (1976) *Working Together: Women And Family in Southwestern Saskatchewan*, Toronto: Rinehart & Winston.

—— (1988) 'Image and behaviour: Women's participation in North American family agricultural enterprises', in W.G. Haney and J.B. Knowles (eds) *Women and Farming: Changing Roles, Changing Structures*, Boulder, Colo.: Westview, pp. 89–108.

Kolodny, A. (1984) *The Land Before Her: Fantasy and Experience of the American Frontier, 1630–1860*, Chapel Hill, NC: University of North Carolina Press.

Lailson, S. (1989) 'La violencia doméstica', *Renglones* 5(15): 60–5.

Langness, L. and Frank, G. (1981) *Lives: An Anthropological Approach to Biography*, Novata, Calif.: Chandler & Sharp.

Lazos Chavero, E. and Alvarez-Buylla Roces, M.E.(1988) 'Ethnobotany in a tropical-humid region: The home gardens of Balzapote, Veracruz, Mexico', *Journal of Ethnobiology* 8 (1): 45–79.

Lefevere, E. and Jackson, K.D. (eds) (1982) 'The art and science of translation', *Dispositio* vii.

Leff, E. (1990) 'Introducción a una vision global de los problemas ambientales de México', in E. Leff (ed.) *Medio Ambiente y Desarrollo en México*, Mexico City: UNAM.

LeGrand, C. (1986) *Frontier Expansion and Peasant Protest in Colombia, 1850–1936*, Albuquerque, NM: University of New Mexico Press.

—— (1989) 'Colonization and violence in Colombia: Perspectives and debates', *Canadian Journal of Latin American and Caribbean Studies* 14 (28): 5–29.

Levi, Y. (1989) 'Relationships between settlers' organizations and the settlement community in new land settlement projects: Cross national experiences', *Land Reform, Land Settlement and Cooperatives* 1–2: 97–121.

Levi, Y. and Naveh, G. (1989) *Towards Self-Management in New Land Settlement Projects*, Boulder, Colo.: Westview.

LeVine, S. in collaboration with Sunderland Correa, C. (1993) *Dolor y Alegría: Women and Social Change in Mexico*, Madison, Wisc.: University of Wisconsin Press.

Lewis, O. (1959) *Five Families*, New York: Basic Books.

Lewis, W.A. (1954) 'Thoughts on land settlement', *Journal of Agricultural Economics* 11: 3–19.

Liebow, E. (1967) *Tally's Corner: A Study of Negro Streetcorner Men*, Boston, Mass.: Little Brown.

Lisansky, J. (1979) 'Women in the Brazilian frontier', *Latinamericanist* 15 (1): 1–3.

—— (1990) *Migrants to Amazonia: Spontaneous Colonization in the Brazilian Frontier*, Boulder, Col.: Westview.

Long, N. and Long, A. (eds) (1992) *Battlefields of Knowledge*, London: Routledge.

Lund, R. (1981) 'Women and development planning in Sri Lanka', *Geografiska Annaler* 63 (B): 95–108.

Lustig, N. (1992) *Mexico: The Remaking of an Economy*, Washington DC: The Brookings Institution.

MacAndrews, C. (1979) 'The role and potential use of land settlements in development policies: Lessons from past experience', *Sociologia Ruralis* 19(2–3): 116–34.

McDowell, L. (1992) 'Doing gender: Feminism, feminists and research methods in human geography', *Transactions of the Institute of British Geographers*, new series 17: 399–416.

McKee, L. (1991) 'Men's rights, women's wrongs: Domestic violence in Ecuador', in D.A. Counts and J.K. Brown (eds) *Sanctions and Sanctuary: Cross-Cultural Perspectives on the Beating of Wives*, Boulder, Colo.: Westview, pp. 139–56.

Madge, C.(1993) 'Boundary disputes: Comments on Sidaway (1992)', *Area* 25 (3): 294–99.

Maos, J.O. (1984) *The Spatial Organization of New Land Settlement in Latin America*, Boulder, Colo.: Westview, Dellplain Latin American Studies No. 15.

Maples, D.E. (1989) 'Building a literary heritage: A study of three generations of pioneer women, 1880–1930', unpublished Ph.D. thesis, University of Missouri-Columbia.

Marín, L. (1991) 'Speaking out together: Testimonials of Latin American women', *Latin American Perspectives* 18(3): 51–68.

Meertens, D. (1988) 'Mujer y colonización en el Guaviare (Colombia)', *Colombia Amazónica* 3(2): 21–56.

——— (1993) 'Women's roles in colonisation: A Colombian case study', in J.H. Momsen and V. Kinnaird (eds) *Different Places, Different Voices: Gender and Development in Africa, Asia and Latin America*, London: Routledge, pp. 256–69.

Merchant, C. (1992) *Radical Ecology: The Search for a Liveable World*, London: Routledge.

Merton, R.K. (1972) 'Insiders and outsiders: A chapter in the sociology of knowledge', *American Journal of Sociology* 78: 9–48.

Mies, M. (1983) 'Feminist research', in G. Bowles and R. Duelli Klein (eds) *Theories of Women's Studies*, London: Routledge.

Molano, A. (1988) 'Violencia y colonización', *Revista Foro* 6: 25–37.

Molino Pineiro, V. and Sanchez Medo, L. (1983) *El Alcoholismo en México*, Mexico City: Fundación de Investigaciones Sociales.

Molloy, S. (1993) *At Face Value. Autobiographical Writing in Spanish America*, Cambridge: Cambridge University Press.

Molyneux, M. (1985) 'Mobilization without emancipation? Women's interests, state and revolution in Nicaragua', *Feminist Studies* 11(2).

Moris, J. (1969) 'The evaluation of settlement schemes performance: A sociological appraisal', in R. Apthorpe (ed.) 'Land settlement and rural development in E. Africa', *Nkanga* 3, pp. 79–102.

Moser, C.O.N. (1989) 'Gender planning in the Third World: Meeting practical and strategic gender needs', *World Development* 17(11): 1799–1825.

——— (1993) *Gender Planning and Development: Theory, Practice and Training*, London: Routledge.

Muench, P. (1982) 'Las regiones agrícolas de Chiapas', *Geografía Agrícola* 2: 57–102.

Murphy, M. (1987) 'The private lives of public women: Prostitution in Butte, Montana, 1878–1917', in S. Armitage and E. Jameson (eds) *The Women's West*, Norman, Okla. and London: University of Oklahoma Press, pp. 193–206.

Naciones Unidas (1989) *Violencia contra la Mujer y la Familia*, New York: United Nations.

Nash, J. and Safa, H. (eds) *Sex and Class in Latin America*, South Hadley, Mass.: Bergin & Garvey.

——— (1986) *Women and Change in Latin America*, South Hadley, Mass.: Bergin & Garvey.

Natera, G. (1983) 'Comparación transcultural de las costumbres y actitudes asociados al consumo de alcohol en dos zonas rurales de Honduras y México', *Revista de Psiquiatria* 29: 116–27.

——— (1987) 'El consumo de alcohol en zonas rurales en México', *Revista de Salud Mental* 10(4): 59–67.

Nations, J.D. and Nigh, R.B. (1980) 'The evolutionary potential of Lacondon Maya sustained-yield tropical forest agriculture', *Journal of Anthropological Research* 36 (1): 1–30

Nelson, M, (1973) *The Development of Tropical Lands*, Baltimore, Md.: Johns Hopkins University Press.

Nelson, N. and Wright, S. (eds) (1994) *Power and Participatory Development: Theory and Practice*, London: Intermediate Technology Publications.

Norwood, V. and Monk, J. (eds) (1987) *The Desert is No Lady: Southwestern Landscapes in Women's Writing and Art*, New Haven , Conn. and London: Yale University Press.

Nugent, S. (1993) 'From 'green hell' to 'green' hell: Amazonia and the sustainability thesis', University of Glasgow, *Amazonian Paper*, no. 3.

Oakley, A. (1981) 'Interviewing women: A contradiction in terms?', in H. Roberts (ed.) *Doing Feminist Research*, London: Routledge.

Oberai, A.S. (1986) 'Land settlement policies and population redistribution in developing countries: Performance, problems and prospects', *International Labour Review* 125(2): 141–61.

—— (1988) 'An overview of settlement policies in developing countries', in A.S. Oberai (ed.) *Land Settlement Policies and Population Redistribution in Developing Countries*, Praeger: New York: Praeger, pp. 7–47.

Ong, A. (1988) 'Colonialism and modernity: Feminist re-presentations of women in non-Western societies', *Inscriptions* 3–4: 79–93.

Oquist, P. (1980) *Violence, Conflict and Politics in Colombia*, New York: Academic Press.

Ordoñez, M. (1986) *Población y Familia Rural en Colombia*, Bogotá: Pontificia Universidad Javeriana.

Palacios, M. (1983) *El Café en Colombia (1850–1970): Una Historia Económica, Social y Política*, 2nd edn, Bogotá: El Ancora.

Palmer, G. (1974) 'The ecology of resettlement schemes', *Human Organization* 33(3): 238–50.

—— (1979) 'The agricultural resettlement scheme: A review of cases and theories', in B. Berdichewsky (ed.) *Anthropology and Social Change in Rural Areas*, The Hague: Mouton, pp. 149–83.

Palmer, I. (1985) *The Impact of Agrarian Reform on Women*, West Hartford, Conn.: Kumarian Press.

Parsons, J.J. (1949) *Antioqueño Colonization in Western Colombia*, Berkeley, Calif.: University of California Press.

—— (1976) 'Forest to pasture: Development or destruction?', *Revista de Biologia Tropical* 24 (suppl. 1): 121–38.

Patai, D. (1991) 'U.S. academics and Third-World women: Is ethical research possible?', in S.B. Gluck and D. Patai (eds) *Women's Words: The Feminist Practice of Oral History*, London: Routledge, pp. 137–54

Patterson-Black, S. (1976) 'Women homesteaders on the Great Plains frontier', *Frontiers* 1: 67–88.

Paz, O. (1950) *El Laberinto de la Soledad*, Mexico City: Cuadernos Mexicanos.

Pelzer, K. (1945) *Pioneer Settlement in the Asiatic Tropics*, New York: Institute of Pacific Relations.

Personal Narratives Group (ed.) (1989) *Interpreting Women's Lives: Feminist Theory and Personal Narratives*, Bloomington and Indianapolis, Ind.: Indiana University Press.

Pescatello, A.M. (1976) *Power and Pawn: The Female in Iberian Families, Societies and Cultures*, London and Westport, Conn.: Greenwood Press

Peters, C.M., Gentry, A.H. and Mendelsohn, R.O. (1989) 'Valuation of an Amazonian rainforest', *Nature* 339: 655–6.

Phillips, L. (1990) 'Rural women in Latin America: Directions for future research', *Latin American Research Review* XXV (3): 89–108.

Pickett, L. (1988) *Organising Development through Participation: Co-operative Organisation and Services for Land Settlement*, London: Croom Helm for the International Labour Office.

Pile, S. (1991) 'Practising interpretive geography', *Transactions of the Institute of British Geographers* 16: 458–69.

Pitt-Rivers, J. (1971) *The People of the Sierra*, Chicago, Ill.: Chicago University Press (first published 1961).

Pontigo Sanchez, J.L. (1990) 'La ganadería bovina en la costa y norte de Chiapas', *Consejo Estatal de Fomento a la Investigación y Difusión de la Cultura* 1(1).

Pratt, M.L. (1992) *Imperial Eyes: Travel Writing and Transculturation*, London: Routledge.

Pulsipher, L. (1993) '"He won't let she stretch she foot": Gender relations in traditional

West Indian houseyards", in C. Katz and J. Monk (eds) *Full Circles: Geographies of Women over the Life Course*, London: Routledge, pp. 107–21.

Radcliffe, S. and Westwood, S. (eds)(1993) *Viva: Women and Popular Protest in Latin America*, London: Routledge.

Rama, A. (1982) *Transculturación narrativa en América Latina*, Mexico City: Siglo XXI.

Randall, M. (1991) 'Reclaiming voices: Notes on a new female practice in journalism', *Latin American Perspectives* 18(3): 103–13.

Reese, L.W. (1991) 'Race, class and culture: Oklahoma women, 1880-1920', unpublished Ph.D. thesis, University of Oklahoma.

Rementería, I. de (1986) 'Hipotesis sobre la violencia reciente en el Magdalena Medio', in G. Sánchez and R. Peñaranda (eds) *Pasado y Presente de la Violencia en Colombia*, Bogota: Fondo Editorial CEREC, pp. 333–48.

Revel-Mouroz, J. (1980) 'Mexican colonization experiences in the humid tropics', in D. Preston (ed.) *Environment, Society and Rural Change*, London: Wiley.

Ribbens, J. (1989) 'Interviewing – An "unnatural situation"?', *Women's Studies International Forum* 12(6): 579–92.

Robles, R., Aranda, J. and Botey, C. (1993) 'La mujer campesina en la época de la modernidad', *Cotidiano* 53: 25–32.

Rocheleau, D.E. (1988) 'Gender, resource management and the rural landscape: Implications for agroforestry and farming systems research', in S.V. Poats, M. Schmink and A. Spring (eds) *Gender Issues in Farming Systems Research*, Boulder, Colo.: Westview.

Roider, W. (1970) *Farm Settlements for Socioeconomic Development: The Western Nigerian Case*, Munich: Weltforum Verlag.

Rose, G. (1993) *Feminism and Geography*, Cambridge: Polity Press.

Rothstein, F. (1986) 'Capitalist industrialization and the increasing cost of children', in J. Nash and H. Safa (eds) *Women and Change in Latin America*, South Hadley, Mass.: Bergin & Garvey.

Rudnick, L. (1987) 'Re-naming the land', in V. Norwood and J. Monk (eds) *The Desert is No Lady: Southwestern Landscapes in Women's Writing and Art*, New Haven, Conn. and London: Yale University Press, pp. 10–26.

Sachs, C.E. (1983) *The Invisible Farmers: Women in Agricultural Production*, Totowa, NJ: Rowman & Allenhead.

Salati, E. (1987) 'The forest and the hydrological cycle', in R.E. Dickinson (ed.) *The Geophysiology of Amazonia*, New York: John Wiley, pp. 273–96.

Salazar, C. (1992) 'A Third World women's text: Between the politics of criticism and cultural politics', in S.M. Gluck and D. Patai (eds) *Women's Words: The Feminist Practice of Oral History*, London:Routledge, pp. 93–106.

Sánchez, G. (1992) 'The Violence: An interpretative synthesis', in C. Bergquist, R. Peñaranda and G. Sánchez (eds)*Violence in Colombia: The Contemporary Crisis in Historical Perspective*, Wilmington: Scholarly Resources Inc., pp. 75–124.

Sánchez, G. and Meertens, D.(1983) *Bandoleros, Gamonales y Campesinos: El Caso de la Violencia en Colombia*, Bogotá: El Ancora.

Sandner, G. (1982) 'El concepto espacial y los sistemas funcionales en la colonización espontanea costaricense', *Revista Geográfica de América Central* 15–16: 95–117.

Saucedo, I. (1993) 'El programa de la identidad de género en la violencia domestica', paper presented to the 13th International Congress of Anthropological and Ethnological Sciences, Mexico City.

Sayer, A. (1984) *Method in Social Science: A Realist Approach*, London: Hutchinson.

Schlissel, L. (1982) *Women's Diaries of the Westward Journey*, New York: Schocken Books.

Schlissel, L., Ruiz, V. and Monk, J. (eds) (1988) *Western Women: Their Land, Their Lives*, Albuquerque, NM: University of New Mexico Press.

Schnetz, M. and Leitzmann, C. (1984) 'Migration and nutrition in Thailand', in H. Uhlig (ed.) *Spontaneous and Planned Settlement in Southeast Asia, Giessener Geographische Schriften*, Hamburg: Institue of Asian Affairs, vol. 58..

Schrijvers, J. (1988) 'Blueprint for undernourishment: The Mahaweli River Development Scheme in Sri Lanka', in B. Agarwal (ed.) *Structures of Patriarchy*, London: Zed Press, pp. 29–51

Scudder, T. (1969) 'Relocation, agricultural intensification and anthropological research', in D. Brokensha with M. Pearsall (eds) *The Anthropology of Development in Sub-Saharan Africa*, The Society for Applied Anthropology, monograph 10, pp. 31–9.

—— (1981) 'The development potential of new land settlement in the tropics and subtropics: A global state of the art evaluation with specific emphasis on policy implications', mimeo, Washington DC: Institute for Development Anthropology Inc. for USAID.

—— (1985) 'A sociological framework for the analysis of new land settlements', in M.M. Cernea (ed.) *Putting People First: Sociological Variables in Rural Development*, 1st edn, New York and Oxford: Oxford University Press for the World Bank, pp. 121–53.

—— (1991) 'A sociological framework for the analysis of new land settlements', revised and expanded, in M.M. Cernea (ed.) *Putting People First: Sociological Variables in Rural Development*, 2nd edn, New York and Oxford: Oxford University Press for the World Bank, pp. 148–87.

Sen, G. and Grown, C. (1987) *Development, Crises and Alternative Visions: Third World Women's Perspectives*, London: Earthscan.

Shiva, V. (1989) *Staying Alive: Women, Ecology and Development*, London: Zed Press.

Simonelli, J.M. (1986) *Two Boys, a Girl, and Enough!* Boulder, Colo.: Westview.

Smith, S. and Watson, J. (eds) (1993) *De/Colonizing the Subject: The Politics of Gender in Women's Autobiography*, Minneapolis, Minn.: University of Minnesota Press.

Solache, G. (1990) 'Encuesta nacional de salud: El consumo de bebidas alcohólicas', *Revista de Salud Mental* 13(3): 13–18.

Sommer, D. (1991) 'Rigoberta's secrets', *Latin American Perspectives* 18(3): 32–50.

Sopher, D. (1983) 'Female migration in Monsoon Asia: Notes from an Indian perspective', *Peasant Studies* 10(4): 289–300.

Southey, S. (1984) 'Women of Mishamo', *Refugees* 2: 11–12.

Spiro, H. (1985) *The Ilora Farm Settlement in Nigeria*, West Hartford, Conn.: Kumarian Press.

—— (1987) 'Women farmers and traders in Oyo State, Nigeria – A case study of their changing roles', in J.H. Momsen and J.G. Townsend (eds) *Geography of Gender in the Third World*, London: Hutchinson, pp. 173–91.

Spivak, G. (1987) *In Other Worlds: Essays in Cultural Politics*, London: Routledge.

—— (1991) 'Neocolonialism and the secret agent of knowledge', *Oxford Literary Review*, special issue on Neocolonialism, 13: 224.

Stacey, J. (1988) 'Can there be a feminist ethnography?', *Women's Studies Forum* 11(1): 21–7; reprinted in S.B. Gluck and D. Patai (eds) (1991) *Women's Words: The Feminist Practice of Oral History*, London: Routledge, pp. 111–20.

Stanley, L. and Morgan, D. (eds) (1993) 'Auto/biography in Sociology', special issue, *Sociology, The Journal of the British Sociological Association* 27(1).

Sternbach, N.S. (1991) 'Re-membering the dead: Latin American women's testimonial discourse', *Latin American Perspectives* 18(3): 91–102.

Stycos, M. (1958) *Familia y Fecundidad en Puerto Rico*, Mexico City: Fondo de Cultura Económica.

Sumarjatiningsih, M.S. (1985) 'Spontaneous transmigration in south Sumatra', *Land Reform, Land Settlement and Cooperatives* 1–2: 75–82.

Taggert, J.M. (1992) 'Gender, segregation and cultural constructions of sexuality in two Hispanic societies', *American Ethnologist* 19(1): 75–96.

Thompson, E.H. (1988) 'The pioneer woman: A Canadian character type', unpublished Ph.D. thesis, University of Western Ontario.

Toledo, V.A. (1990) 'El proceso de ganaderización y la destrucción biológica e ecológica en México', in E. Leff (ed.) *Medio Ambiente y Desarrollo en México*, Mexico City: UNAM.

Townsend, J. (1976) 'Land and society in the middle Magdalena valley, Colombia', unpublished D.Phil. thesis, Oxford.

—— (1977) 'Perceived worlds of the colonists of tropical rainforest, Colombia', *Transactions of the Institute of British Geographers*, new series 2(2): 430-58.

—— (1985) 'Seasonality and capitalist penetration in the Amazon Basin', in J. Hemming (ed.) *Change in the Amazon Basin*, vol. 2: *The Frontier after a Decade of Colonisation*, Manchester: Manchester University Press, pp. 140–57.

—— (1989) 'Cooperativa Integral de la Comuna de Payoa Ltda', *Cuadernos de Agroindustria y Economía Rural* 20: 71–0.

—— (1991) 'Geografía y género en la colonización agrícola', *Documents d'Anàlisi Geogràfica* 18: 89–99.

—— (1993a) 'Gender and the life course on the frontiers of settlement in Colombia', in C. Katz and J. Monk (eds) *Full Circles: Geographies of Women over the Life Course*, London: Routledge, pp. 138–55.

—— (1993b) 'Housewifization and colonisation in the Colombian rainforest', in J.H. Momsen and V. Kinnaird (eds) *Different Places, Different Voices: Gender and Development in Africa, Asia and Latin America*, London: Routledge, pp. 270–7.

—— (1994) 'Who speaks for whom? Outsiders represent women pioneers of the forests of Mexico', in N. Nelson and S. Wright (eds) *Power and Participatory Development: Theory and Practice*, London: Intermediate Technology Publications.

Townsend, J., Arrevillaga, U., Cancino, S., Pacheco, S. and Pérez, E. (1994) *Voces Femeninas de las Selvas*, Montecillo, Mexico: Colegio de Postgraduados and University of Durham.

Townsend, J. with Bain de Corcuera, J. (1993) 'Feminists in the rainforest in Mexico', *Geoforum* 24(1): 49–54.

Townsend, J. and Wilson de Acosta, S. (1987) 'Gender roles in the colonization of rainforest: A Colombian case study', in J. Momsen and J.G. Townsend (eds) *Geography of Gender in the Third World*, London: Hutchinson.

Tudela, F. (co-ordinator) (1989) *La Modernización Forzada del Tropico: El Caso de Tabasco*, Mexico City: El Colegio de México.

Turner, F.J. (1921) *The Frontier in American History*, New York: H. Holt & Co.

Uhlig, H. (1984) 'Government-sponsored versus spontaneous settlement?', in H. Uhlig (ed.) *Spontaneous and Planned Settlement in SouthEast Asia, Giessener Geographische Schriften*, Hamburg: Institute of Asian Affairs, vol. 58, pp. 105–18

Ulluwishewa, R. (1989) 'Development planning and gender inequality: A case study in Mahaweli Development Project in Sri Lanka', paper presented at the Commonwealth Geographical Bureau Workshop on Gender and Development, Newcastle upon Tyne.

Uquillas, J.E. (1989) 'Social impacts of modernization and public policy and prospects for indigenous development in Ecuador's Amazonia', in D.A. Schumann and W.L. Partridge (eds) *Human Ecology of Tropical Land Settlement in Latin America*, Boulder, Colo.: Westview Press, pp. 407–31.

Valentine, G. (1989) 'The geography of women's fear', *Area* 21(4): 385–90.

Van Kirk, S. (1987) 'The role of native women in the creation of fur trade society in Western Canada, 1670–1830', in S. Armitage and E. Jameson (eds) *The Women's West*, Norman, Okla. and London: University of Oklahoma Press, pp. 53–62.

van Raay, H.G.T. and Hilhorst, J.G.M. (1981) *Land Settlement and Regional Development in the Tropics: Results, Prospects and Options*, The Hague: ISS Advisory Service.

Vayda, A.P. (1987) 'Self-managed land colonisation in Indonesia', in D.C. Korten (ed) *Community Management: Asian Experience and Perspectives*, West Hartford, Conn.: Kumarian Press.

Velasco Muñoz, M. del P. (1983) 'Causas sociales del alcoholismo en México', in *Memorias del Seminario de Análisis sobre el Alcoholismo en México*, conference paper, Mexico City: Fundación de Investigactiones Sociales.

Weil, J.E. (1980) 'The organisation of work in a Quechua pioneer settlement: Adaptation of highland traditions to the lowlands of Eastern Bolivia', unpublished Ph.D. dissertation, New York: Colombia University Press

Weitz, R. (1971) *From Peasant to Farmer: A Revolutionary Strategy for Development*, New York: Colombia University Press.

Weitz, R., Pelley, D. and Applebaum, L. (1980) 'A model for the planning of new settlement projects', *World Development* 8: 705–23.

Welter, B. (1966) 'The Cult of True Womanhood: 1820-1860', *American Quarterly* 18(2): 151-74.

Wilson, F. (1991) *Sweaters: Gender, Class and Workshop-Based Identity in Mexico*, Basingstoke: Macmillan.

World Bank (1985) 'The experience of the World Bank with government-sponsored land settlement', unpublished report no. 5625, Washington DC: World Bank.

—————— (1987) *The Jengka Triangle Projects in Malaysia*, Washington DC: The World Bank.

—————— (1991) *World Development Report*, New York: Oxford University Press.

Yúdice, G. (1991) '*Testimonio* and postmodernism: Whom does testimonial writing represent?' *Latin American Perspectives* 18(3): 15–31.

Yuvajita, P. (1986) 'The changing images of women in Western American literature as illustrated in the works of Willa Cather, Hamlin Garland and Mari Sandoz', unpublished Ph.D. thesis, University of Oregon.

Zapata, E.(1990) *Sueños y Realidades de la Mujer*, Montecillo, Mexico: Colegio de Postgraduados.

Zeballos-Hurtado, H. (1975) 'From the uplands to the lowlands: An economic analysis of Bolivian rural–rural migration', unpublished doctoral dissertation, University of Wisconsin.

INDEX

academics as experts? 8, 10, 58, 131, 132, 133, 140
access to markets and services 37–40, 67–9, 70–3, 84, 124
advice: to daughters 95–7, 173, 175, 189; to other women 98–9, 166, 176, 192; to sons 97, 173, 175, 189–90
Agricultural and Industrial Women's Group (UAIM) 56, 120–1, 127, 130, 161–5, 173
agriculture: land settlement 21–33; ownership of farms 20; women's work in 10, 20, 28–9, 43–4, 87, 100–1, 120–1, 160, 180; *see also* cattle; gardens
agroforestry 62–3, 77
alcoholism 57, 66, 73, 77, 109–11, 124–5, 152, 178–9; banning of alcohol 66, 67, 116, 126, 130; and marital rape 107–8, 110; problems associated with 71; rates of 111; and violence 114, 130, 169; in women 111
Alsop, R. 49
Alvarez-Buylla Roces, M.E., Lazos Chavero, E. and Garcia-Barrios, J.R. 61
Amazonia 27, 33; TransAmazon Highway 22, 28; use of new land 22
Apthorpe, R. 25
Arizpe, L. and Botey, C. 120, 127, 128
Arizpe, L., Botey, C., Salinas, F. and Velásquez, M. 101
Arizpe, L., Paz, F. and Velásquez, M. 74, 76, 81, 106
Armitage, S. 20
Armitage, S. and Jameson, E. 19
Armstrong, A. 32
Arndt, H.W. 24
Arrevillaga, Ursula 13, 57

Bahrin, T.S., Thong, L.B. and Dorall, R.F. 32
Bain, J. 3, 12, 56–7
Basnett, S. and Lefevere, A. 137
Basnett-McGuire, S. 138, 139
Basso, E. 10, 139, 141
Beenstock, M. 25
begging 84, 101
Belshaw, D.G.R. 25
Benería, L. and Roldán, M. 129
biodiversity 123–4
Bolivia 48; women and land settlement schemes 32
Boserup, Ester 21
Boulding, E. 123
Bowman, Isaiah: *The Pioneer Fringe* 18
Brazil 35; TransAmazon Highway 22, 28; Tucumá project 29
Brown, E. 113, 114
Brunt, Dorien 58
Bunyard, P. 26
Burkina Faso 29
Butler, J.R. 25, 29

Cameron, D. 138
Campbell, J.C. 114
Cancino, Socorro 13
Cather, Willa 19
Catholic Church 120, 152, 186–7; and land settlement 39–40; Liberation Theology 127
cattle raising: in Colombia 32, 35; in Mexico 59, 60, 65, 66, 71; women's work in 32, 43–4, 65, 165
Cernea, M. 22, 25
Chambers, R. 9, 11, 24, 25, 28, 29–30, 43, 123, 133
Chambers, R., Pacey, A. and Thrupp, L. 2

206

Chant, S. 94, 113
chaperonage 89–90, 104
child abuse 104, 109
childcare 29; sharing of 30
children: advice to 95–7, 173, 175, 189–90; attitudes to 93–5, 150, 170; kept out of forest 83; nutrition of 38–9, 48; opportunities for play 155, 167–8; punishment of 154, 155–6, 161, 167, 181; relationships with parents 71, 93, 148, 150, 167, 168–9, 181, 184–5; value placed on 100, 108
China: Three Gorges Dam 22
Clifford: James 15
co-operation: breakdowns in 161–5; by communities 125–6; by women's groups 127; necessity of 85; and self-help 121; support for 132
co-operatives 30; La Payoa settlement 39–40, 44–5, 48–9
Coberly, R. 93
coffee growing 35, 180
Colombia 26, 29; cattle rearing 32, 35; drug industry 35, 38; El Distrito settlement 39, 46, 48; gender roles 41–3; guerrilla groups 35–6; health care 37, 46–7; high-tech farming 39; La Payoa settlement 39–40, 44–5, 48–9; Magdalena Medio valley 1, 2, 3, 36–49, 124; migration to lowlands 34, 35–6, 47; nutrition in 38–9, 48; paramilitary squads 36–7; San Lucas settlement 37–9, 42, 44, 45–6, 48; sex ratios in settlements 40–1, 48; social isolation in 30; 'the Violence' 35–6; title to land 43; violence in 34–6, 39–40; women defined by family 42; women's work with cattle 32; women's work on land 10
Colombian Federation of Coffee Growers 35
colonialism: and land settlement 21, 24
colonisation: reasons for 81–3; see also land settlement
Colson, E. 25
communication, and issues of translation 10, 137–43
communities: call for creation of public spaces 126, 192; effectiveness of leaders 125; feuds within 125; framework provided by 125; insiders' wishes for improvements 126, 131–3, 192–3;

male-oriented priorities 125; manipulation of outside agencies 16, 57–8
contraception 102, 106–8, 130, 191; attitudes to 61, 73; resistance to 107–8
cooking 20, 44, 63, 84–5
Cotterill, P. 16
Court, G. 43
Cubitt, T. 94
Currier, R. 89

dams and land settlement 22
dances 168–9
Davies, C.B. 133
debt 165; international 132
deconstruction and translation 13, 138–9
Deere, C.D. 29
Deere, C.D. and Leon de Leal, M. 41, 42, 43, 47
deforestation 27, 65, 184; attitudes to 74–7, 170–1
Derrida, Jacques 138
development: 'deteriorating' 50–1, 60
development studies, male orientation of 133
Diaz Guerrero, R. 94
displacement, pioneering and 7
domestic work see work
Douglas, C.B. 91
Dozier, C.L. 24
drugs 1, 175; Colombian drug industry 35, 38

ecofeminism 74, 76
Ecuador 26
education 145; absences of teachers 118, 126, 130; attitudes to 86–7, 192; belief in power of 58, 132, 175–6; costs of 151, 167, 171, 174–5; deficiencies of training programmes 127–8, 130; effect on women's lives 85–6; levels of 38–40, 58, 60, 73, 86; opportunities for 60, 73, 85–7, 118, 180–1; role of community workers 128–9; sacrifices made for 86–7; telesecondary schools 118, 151; training projects for women 71–2, 119, 127–8, 130; women re-entering 174; see also literacy
ejido (land reform community) 52, 53, 56, 59, 60, 68, 81–2, 162; failures of 189–90; formation 72; male dominance of 99; problems of 177–8

emic and etic views 4, 8, 10, 25, 49, 70, 131
entrepreneurs 102–3, 145, 147–8, 152–3, 159–60, 171
environmental damage 26–7, 35, 51, 74–6
Ewell, P.T. and Poleman, T.T. 26, 52, 70
exploitation in research 16, 57–8

families: age at marriage 65, 67, 73, 104; and attitudes to contraception 61; effects of agrarian crisis on 60–1; in Mexico 59; parent-child relationships 71, 93, 148, 150, 167, 168–9, 181, 184–5; power of husband in nuclear family 33; relationships with grandparents 88; seclusion of girls 89–91, 104; single parent 99–100; size of 38, 61; step-parents 92–3; support services for 68; supported by women 101–2; women defined by 42; women's separation from support of 71, 124
Family Integration, Department for (DIF) 68, 71, 130–1
Feder, E. 60
femininity 20, 42, 93–7, 102, 117, 151–2
feminism 11; analysis of poverty 123; attitudes to environment 74, 76; and issues of translation 137–43
feminist methodology 14–17; interviewing 15–16; life stories 16–17; literature review 15; see also interviewing techniques
Fernandez Serra, M.T. 25
fertilisers 60, 62
fertility, women's 38–40, 60–1, 106–8
freedom of movement: seclusion of girls 89–91; of women 66, 173–4
Frenk, Susan F. 13
Frick, M.J.B. 19
friendship 117–18, 155, 163, 173- 4, 186
Fromm, E. and Maccoby, M. 94

gardens 73, 76–7, 95, 102, 112; and cash economy 62; family work in 59, 61–2; growing of medicinal plants 89; teaching about in schools 119–20
Garlin, Hamlin 19–20
gender: and attitudes to deforestation 74–7
gender expectations 93–4; attitudes to being a woman 94–5

gender roles: in Colombia 41–3; 'Cult of True Womanhood' 20; (tables) 42, 63–7
Ghana, Volta Dam 22
Gluck, S.B. 17
Gonzalez de la Rocha, M. 113
Gonzalez Montez, S. and Iracheta Cenegorta, P. 114
Gosling, L.A.P. and Abdullah, H. 25
gossip 85, 117, 155, 163, 174
Greenberg, J.B. 130
guerrilla groups, Colombia 35–6
Guggenheim, S.E. 25
Guinness, P. 25

Hahn, N.D. 29
Hamilton, S. 32, 48
Hanger, J. and Moris, J. 31
Haraway, Donna 9, 15
Harding, S. 133
Harris, K. 19, 21
health: in Colombia 37, 46–7; costs of illness 147, 151; growing of medicinal plants 62, 63; and hardships of early communities 83, 84; infestation by parasites 65; primary health care needs 130; reproductive problems 191; role of health committees 126; role of health workers 71–2; and sanitary facilities 65; stress-related illness 148–9; women's anxieties about 141–2; see also childbirth; contraception; sterilisation
Hecht, S.B. 32, 43
Hecht, S.B. and Cockburn, A. 27, 33, 35
Higgs, J. 25
Hirschmann, A.O. 23
home 42, 63, 77, 84, 86, 94–5, 118, 144, 146
homeworking 129, 131–2
homosexuality 42, 104
hospitals, access to 84; see also health
Hulme, David 8, 23, 26, 133

income: earnings levels 119, 128, 129; levels of pooling 29, 30–1; of women in land settlement schemes 31, 32, 43–5, 63–4, 100-3
India: Narmada Valley Dams 22; violence towards women 114
Indonesia 24, 26; research on settlers 26; use of new land 22; women's financial training 30
infidelity 108–9, 157

insiders 4, 8, 14, 19, 25–6, 131–2, 144–93; improvements wanted by 126, 131–3, 192–3

interviewing techniques 11, 15–16, 38, 48–9; inclusion of women 28; and interpretation of oral exchanges 141–2; leading questions 140; treatment of repetition 141–2; workshops 55–6, 73, 94–5

invisibility of pioneer women 9

isolation 21, 30, 31, 45–6, 59, 70, 158, 180

Jackson, C. 28

Jacobs, S. 28, 29, 32, 123

James, W.E. 25

Jameson, E. 19, 20, 21

Jeffery, J.R. 19

Jensen, J.M. and Miller, D.A. 19, 41

Joly, L.G. 26

Jones, J.J. 21, 27, 33

Katzman, M.T. 25

Kearney, R.N. and Miller, B.D. 41

Kedar, L. 140

Kenya, Mwea Scheme 31

Kohl, S.B. 19, 20, 21

Lailson, S. 114

land reform programmes: Colombia 35, 39, 48; Mexico 51, 52; women's problems with 39

land settlement 8, 21–33; and colonialism 21; costs of 25–6, 32, 33; effects on women 123; and family relationships 31, 32, 33; feasibility studies 26; future of 132–3; and inequality 23; and isolation 180; literature on 24; and natural disasters 22; and nomads 22; perceptions of change 187–8; and poverty 82, 123; quality of administrators 24–5; reasons for 22–3, 34–6, 51, 81–3, 123; and refugees 22, 32; sex ratios 40–1; social considerations 24; and social isolation 30, 45–6, 180; spontaneous 8, 25, 26, 48, 51, 59; study of women's role 27–9; success in 23–6; sustainable 61–5, 123–4; women's problems in 29–33; and women's work 29–30

land speculation 22, 27

Langness, L. and Frank, G. 142

language, dialects 185

laundry 20, 45, 63

Lazos Chavero, M.E. and Alvarez-Buylla Roces, M.E. 61, 62

Lefevere, E. and Jackson, K.D. 142

LeGrand, C. 35, 36

Lehmann, D. 132

leisure 100

Levi, Y. 25

Levi, Y. and Naveh, G. 25

LeVine, S. with Sunderland Correa, C. 85–6, 87, 88, 89, 91, 93, 94, 106, 108, 114

Lewis, O. 93

Lewis, W.A. 24

liberation, women's achievement of 116

life stories 11, 16–17, 55, 139–43, 144–93

Lisansky, J. 32, 43–4, 48

literacy 60; gained by adults 86; Mexico 73, 86

literature: on land settlement 24; of pioneer women 19–20

Long, N. and Long, A. 58

Lund, R. 30, 31

MacAndrews, C. 25

McDowell, L. 15

machismo 107; and alcoholism 111; and attitudes to women's work 119

Madge, C. 11

Malaysia 29; FELDA schemes 28; Jengka Triangle Projects 31–2; land settlement schemes 26, 32

male-biased research 2, 3, 4, 42

manipulation, and life stories 17, 57–8

Maos, J.O. 25

marketing: future approaches to 129; of labour-intensive products 63–5

marriage 45, 146; age at 65, 67, 73, 104; attitudes to 149; coercion into 90; and control over money 111–12; expectations of breakdown 96, 104; and infidelity 108–9, 157; polygamy 182; and separation 182; and sexuality 105, 191; women's evaluations of 117, 157–8, 160–2, 168–70, 172–3

masculinity 42, 91, 93–4, 97, 111, 114

Meertens, D. 32, 41, 42, 43, 48

menopause 105, 186, 190–1

menstruation 55, 156–7, 168, 181–2; attitudes to 88–9

Merchant, C. 62

Merton, R.K. 8

methods 2–4, 8–17, 37, 48–9, 52–7, 81, 122, 125, 137–43

Mexico 10; agrarian crisis 60–1; attitudes to deforestation 74–7; Campeche settlements 70–3; cattle raising 59, 60, 65, 71; 'deteriorating development' 50–1, 60; family feuds 68; family structures 59; forms of land tenure 53, 81; land reform programmes 51, 52, 53; literacy 73, 86; Los Tuxtlas region 61–6, 144; Marques de Comillas area 53; nutrition 73; Oaxaca and Veracruz settlements 67; Palenque area 68; Plan Balancán-Tenosique 68–70; political system 58–9, 124; pollution of Mexico City 76; poverty in 70; proposals for changes in national policy 131; research methods 54–7; rights to land 51, 52; social life of women 66; socialism in 12–13, 15; south-east lowlands 50–2; specific problems of pioneers 124; title to land 59; Uxpanapa rainforests 26–7, 52

midwifery 20, 101, 159, 188–9

Mies, Maria 15, 16

migration: in Colombia 47; of girls to urban areas 40–1, 45, 61; reasons for 81–3

Molano, A. 35, 43

Molloy, S. 140

Molyneux, M. 49

money, control over within marriage 111–12

Moris, J. 25

Moser, C. 49, 125

mothers, relationships with children 71, 93, 148, 150, 167, 168–9, 181, 184–5

Nash, J. and Safa, H. 94

Natera, G. 111

Nations, J.D. and Nigh, R.B. 50

natural disasters, and land settlement 22

neighbours, relationships with 85, 117, 155, 163, 174

Nelson, M. 25

Nelson, N. and Wright, S. 9

Nestlé 65

Nigeria 29; Ilora Farm Settlement 28, 30–1; women's role in agriculture 28

nomads, and land settlement 22

non-governmental organizations (NGOs) 129

Norwood, V. and Monk, J. 19

nutrition: in Colombia 38–9, 48; in Mexico 73

Oberai, A.S. 23, 24

Oquist, Paul 35

Ordoñez, M. 41

organic foods 63–5

oustees 7

outsiders 4, 8, 9, 10, 14, 19, 34, 62, 123–31

ownership of farms 20

Oxfam, project in El Tulipán (Mexico) 70, 124

Pacheco, Silvana 13, 55, 57

Palacios, M. 35

Palmer, G. 24

Palmer, I. 29, 30

Panama 26

paramilitary squads, Colombia 1, 36–7

parent-child relationships 71, 93, 148, 150, 167, 168–9, 181, 184–5

Parsons, J.J. 60; *Antioqueño Colonization in Western Colombia* 35

Patai, D. 15, 16

Paz, O. 94

Pelzer, K. 24

Pérez, Elia 13

Personal Narratives Group 142

Pescatello, A.M. 91, 94

pesticides 62, 65; contamination by 47

Peters, C.M., Gentry, A.H. and Mendelsohn, R.O. 27

Phillips, L. 17

Pickett, L. 30

pioneering: seen as answer to rural poverty 8, 21

pioneers: definition 1, 7; improvements wanted by 126, 131–3, 192–3

Pitt-Rivers, J. 90, 91

poverty: gender differences in 123; and land settlement 21, 123; in Mexico 70; as motive for pioneer women 20, 81–2, 183–4

power relations 11; difficulties in researching 48–9

practical and strategic needs 49, 118–21, 125–9

Pratt, M.L. 139

pregnancy and childbirth: experience of 158–9, 162, 183; lack of help in 30, 46, 106; role of midwives 101, 188–9

private sector 129
propriety, demanded of women 86, 151–2, 182
prostitution 32, 45, 108–9
public roles of women 103
public spaces, women's desire for 126, 192

questionnaires 10, 42, 48, 54

race, as social category 11–12, 18–21
Radcliffe, S. and Westwood, S. 94
Rama, A. 139
rape, marital 57, 107–8, 109, 124–5
Reese, L.W. 19
refugees 22, 32
religion: and collective action 127; women's groups based on 120
Rementería, I. de 36
repertory grid techniques 2, 3
representation, by academic 'outsiders' 8, 15, 58, 139–140
research project: male support for 58; manipulation of by women 17, 57–8
rice crops 39, 71
Robles, R., Aranda, J. and Botey, C. 120
Roider, W. 30, 31
Rose, G. 133
Rothstein, F. 106

safety: standards of 73; see also violence
Salati, E. 27
Salazar, C. 17
Sandner, G. 26
Sandoz, Mari 19–20
Saucedo, I. 108
Sayer, A. 50
Schentz, M. and Leitzmann, C. 30
Schlissel, L., Ruiz, V. and Monk, J. 19, 20
Schrijvers, J. 28, 29, 31
Scudder, T. 23, 25, 28, 30, 31, 123, 132
seclusion 77, 89–90, 117–18
Sen, G. and Grown, C. 123
Seventh Day Adventists 120, 127, 152
sex ratios: in Colombian settlements 40–1, 48; and division of labour 63–4; in Mexican settlements 59
sexual abuse 92–3; see also rape
sexuality 42, 45, 55, 178–9, 191; attitudes to 105; and conflicts around reproduction 57; and marriage 105, 191; and seclusion of girls 89–91, 104; women's attitudes to 96–7

sexual relations 104–5, 117, 178
Shiva, Vandana 74–5
Simonelli, J.M. 106
Smith, S. and Watson, J. 140
social isolation 30, 45–6, 180
social life: attitudes to talking with other women 70; of men 161, 172; television 70; of women 21, 66, 163, 186; women's friendships 117–18
socialism, in Mexico 12–13, 15
Solache, G. 111
solutions 62–3, 120–2, 125–31
Sopher, D. 41
Southey, S. 32
Spain, influence of in Latin America 90–1
Spiro, H. 28, 29, 30, 31
Spivak, Gayatri Chakravorty 139
spontaneous settlement 8, 25, 26, 48, 51, 59
Sri Lanka 29, 33; Dry Zone schemes 41; Mahaweli River Development Scheme 22, 28, 31; social isolation in 30
Stacey, J. 11, 16
standpoint theories 133
Stanley, L. and Morgan, D. 140
state, recommendations to 129–31
State Department of Family Integration (DIF) 68, 71, 130–1
sterilisation 106–8, 130, 147–8, 150, 159, 174–5; male and female attitudes to 106–8
Stycos, M. 90
subaltern voices 13, 139
Sudan: women's co-operatives 30
suffering, womanhood affirmed in terms of 117
Sumarjatiningsih, M.S. 25
sustainable settlement 26–7; measures necessary for 124

Taggert, J.M. 91
Tanzania: Mishamo settlement 32; settlement of refugees 22
teachers: absences of 118, 126, 130
technology: belief in 2, 26, 51, 71; transfer of 2
television: influence on attitudes 76; role of soap operas 70
Thailand 26; Chonburi Hinterland 30, 33
title to land 21, 29, 30–1, 32; Colombia 43; Mexico 59
Toledo, V.A. 60

town planning 69–70, 72
Townsend, J. 1–4, 9, 12, 27, 29, 30, 35, 42
Townsend, J. with Bain, J. 56–7, 67, 106
Townsend, J. and Wilson de Acosta, S. 3,
    9, 21, 32, 38, 41–2
trainers 71, 119, 127–8
transculturation 10, 14, 138–9, 143
translation issues 137–43
Tudela, F. *et al.* 50–1, 60, 68
Turner, Frederick Jackson 19

UAIM (Agricultural and Industrial
    Women's Group) 56, 120–1, 130,
    161–5, 173; weaknesses of 127
Uhlig, Harald 21, 26, 30, 33
Ulluwishewa, R. 29, 30, 31
United States: pioneer women in 19–20,
    33
Uquillas, J.E. 26
USAID 28

Valentine, G. 112
Van Kirk, S. 19
van Raay, H.G.T. and Hilhorst, J.G.M. 25
Vayda, A.P. 25
Velasco Munoz, M. del P. 111
violence: against children 91–2, 154,
    155–6, 161, 167, 181; and alcoholism
    114, 130; in Colombia 34–6, 39–40;
    differences between communities
    113–14; domestic 19, 48, 57, 71, 77,
    90–3, 98–9, 108, 112–16, 124–5, 131,
    161, 169, 172, 183; in Mexico 68;
    reasons for 114; in settlement areas
    34–6, 65, 70, 82, 144–5; women's
    resistance to 115–16

water: in Campeche settlements (Mexico)
    70; collection of 49, 63, 71, 72–3, 125,
    186; and sanitary facilities 65, 72
Weil, J. 28–9
Weitz, R. 25
Welter, B. 20
widowhood 99
Wilson, F. 129, 132
Wilson de Acosta, Sally 3, 10, 12
women's groups 120–2, 186; and
    collective action 127; discussion of
    gender needs 121–2; enthusiasm for
    132; UAIM 56, 120–1, 127, 130,
    161–5, 173
work: attitudes to women's work 20, 61,
    95, 98, 118–20; by girls 87–8, 150, 154,
    155, 180; collective sewing scheme
    184–5; division of labour by gender 42–
    7, 59, 63–4; domestic 45, 46–7, 88,
    171; domestic work by girls 88, 150,
    154, 167, 180; food preparation 84–5;
    improvements in types and conditions
    of 127–31; off-farm work 101–2;
    opportunities for urban women 61, 66,
    67; opportunities for women 61; paid
    domestic work 45, 61; selling goods
    102–3, 145, 171–2; women
    entrepreneurs 102–3, 152–3, 158–9;
    women's work in agriculture 10, 28–9,
    41, 43–4, 87, 100–1, 120–1, 147, 160,
    165, 180; workload of women 29–30,
    45
workshops 55–6, 73, 94–5, 128
World Bank 7, 22, 23, 24, 25, 28, 29,
    31–2, 123

Yúdice, G. 15

Zapata, E. 128
Zeballos-Hurtado, H. 25
Zimbabwe 32